Automotive Engineering: An Introduction

Automotive Engineering: An Introduction

Edited by Vincent Parese

Clanrye International
www.clanryeinternational.com

Clanrye International,
750 Third Avenue, 9th Floor,
New York, NY 10017, USA

ISBN: 978-1-63240-710-8

Cataloging-in-Publication Data

Automotive engineering : an introduction / edited by Vincent Parese.
p. cm.
Includes bibliographical references and index.
ISBN 978-1-63240-710-8
1. Automobiles--Design and construction. 2. Motor vehicles--Design and construction.
I. Parese, Vincent.
TL240 .A98 2018
629.231--dc23

For information on all Clanrye International publications visit our website at www.clanryeinternational.com

CLANRYE INTERNATIONAL

Printed in China.

Contents

Preface

The study and practice of designing, constructing, manufacturing and operating automobiles is known as automotive engineering. It is a sub-field of vehicle engineering. It is based on the elements of software engineering, electrical engineering, safety engineering and mechanical engineering, etc. The subject has three main parts namely designing the different aspects of a vehicle, testing these parts, and final manufacturing. This book is a compilation of chapters that discuss the most vital concepts in the field of automotive engineering. Such selected concepts that redefine the area have been presented in it. For all those who are interested in automotive engineering, this textbook can prove to be an essential guide.

A detailed account of the significant topics covered in this book is provided below:

Chapter 1- Automobile engineering is a branch of engineering which is mainly concerned with the design and manufacturing of automobiles and motorcycles. This subject has three main functions, production, manufacturing and development. This chapter is an overview of the subject matter incorporating all the major aspects of automotive engineering.

Chapter 2- The development of the physical appearance and functionality of road vehicles is known as automotive design. Some of the important disciplines related to this subject are safety engineering, vehicle dynamics, ergonomics engineering, etc. This chapter will provide an integrated understanding of automotive designing.

Chapter 3- Motorcycle design is the art of designing and engineering motorcycles. Motorcycle braking systems, motorcycle fork, shock absorber, hub-center steering, sidecar, featherbed frame and motorcycle fairing has been explained. The chapter on motorcycle design and technologies offers an insightful focus, keeping in mind the complex subject matter.

Chapter 4- Custom cars are cars that are made as per the requirements and choice of the driver. The machine is altered in order to improve its performance and to make it look different. The features added are paint, engines swapping, etc. This chapter is a compilation of the various automotive styling features that form an integral part of the broader subject matter.

Chapter 5- Some of the essential aspects of automotive engineering such as body design and essential components that keep the body stable, supported as well as boost ergonomic features of the vehicle. The section includes topics like H-point, Multi leaf spring, etc. Automotive engineering is best understood in confluence with the major topics listed in the following chapter.

It gives me an immense pleasure to thank our entire team for their efforts. Finally in the end, I would like to thank my family and colleagues who have been a great source of inspiration and support.

Editor

1

An Introduction to Automotive Engineering

Automobile engineering is a branch of engineering which is mainly concerned with the design and manufacturing of automobiles and motorcycles. This subject has three main functions, production, manufacturing and development. This chapter is an overview of the subject matter incorporating all the major aspects of automotive engineering.

Automotive Engineering

Automobile engineering, along with aerospace engineering and marine engineering, is a branch of vehicle engineering, incorporating elements of mechanical, electrical, electronic, software and safety engineering as applied to the design, manufacture and operation of motorcycles, automobiles and trucks and their respective engineering subsystems. It also includes modification of vehicles. Manufacturing domain deals with the creation and assembling the whole parts of automobiles is also included in it.The automotive engineering field is research -intensive and involves direct application of mathematical models and formulas. The study of automotive engineering is to design, develop, fabricate, and testing vehicles or vehicle components from the concept stage to production stage. Production, development, and manufacturing are the three major functions in this field.

Automobile Engineering is mainly divided into three streams such as production or design engineering focuses on design components, testing of parts, coordinating tests, and system of a vehicle.

Disciplines

Automobile Engineering

Automobile Engineering is a branch study of engineering which teaches manufacturing, designing, mechanical mechanisms as well operations of automobiles. It is an introduction to vehicle engineering which deals with motorcycles, cars, buses trucks etc. It includes branch study of mechanical, electronic, software and safety elements. Some of the engineering attributes and disciplines that are of importance to the automotive engineer and many of the other aspects are included in it.

Safety engineering: Safety engineering is the assessment of various crash scenarios and their impact on the vehicle occupants. These are tested against very stringent governmental regulations. Some of these requirements include: seat belt and air bag functionality testing, front and side impact testing, and tests of rollover resistance. Assessments are done with various methods and tools, including Computer crash simulation (typically finite element analysis), crash test dummies, and partial system sled and full vehicle crashes.

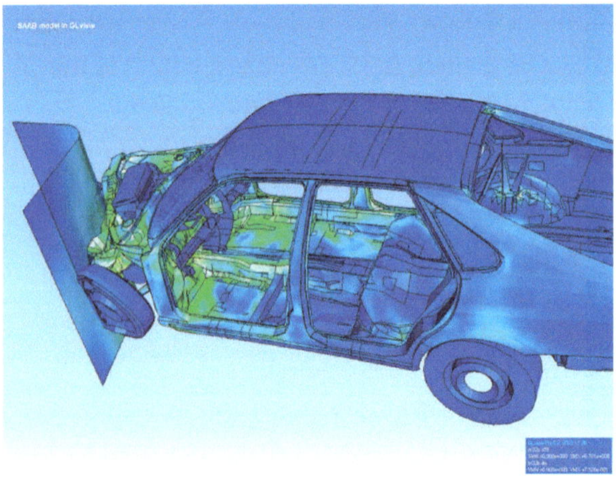

Visualization of how a car deforms in an asymmetrical crash using finite element analysis.

Fuel economy/emissions: Fuel economy is the measured fuel efficiency of the vehicle in miles per gallon or kilometers per liter. Emissions testing includes the measurement of vehicle emissions, including hydrocarbons, nitrogen oxides (NOx), carbon monoxide (CO), carbon dioxide (CO_2), and evaporative emissions.

Vehicle dynamics: Vehicle dynamics is the vehicle's response of the following attributes: ride, handling, steering, braking, comfort and traction. The design of the chassis systems of suspension, steering, braking, structure (frame), wheels and tires, and traction control are highly leveraged by the vehicle dynamics engineer to deliver the vehicle dynamics qualities desired.

NVH engineering (noise, vibration, and harshness): NVH is the customer's feedback (both tactile [felt] and audible [heard]) from the vehicle. While sound can be interpreted as a rattle, squeal, or hot, a tactile response can be seat vibration or a buzz in the steering wheel. This feedback is generated by components either rubbing, vibrating, or rotating. NVH response can be classified in various ways: powertrain NVH, road noise, wind noise, component noise, and squeak and rattle. Note, there are both good and bad NVH qualities. The NVH engineer works to either eliminate bad NVH or change the "bad NVH" too good (i.e., exhaust tones).

Vehicle Electronics: Automotive electronics is an increasingly important aspect of automotive engineering. Modern vehicles employ dozens of electronic systems. These systems are responsible for operational controls such as the throttle, brake and steering controls; as well as many comfort and convenience systems such as the HVAC, infotainment, and lighting systems. It would not be possible for automobiles to meet modern safety and fuel economy requirements without electronic controls.

Performance: Performance is a measurable and testable value of a vehicle's ability to perform in various conditions. Performance can be considered in a wide variety of tasks, but it's generally associated with how quickly a car can accelerate (e.g. standing start 1/4 mile elapsed time, 0–60 mph, etc.), its top speed, how short and quickly a car can come to a complete stop from a set speed (e.g. 70-0 mph), how much g-force a car can generate without losing grip, recorded lap times, cornering speed, brake fade, etc. Performance can also reflect the amount of control in inclement weather (snow, ice, rain).

Shift quality: Shift quality is the driver's perception of the vehicle to an automatic transmission shift event. This is influenced by the powertrain (engine, transmission), and the vehicle (driveline, suspension, engine and powertrain mounts, etc.) Shift feel is both a tactile (felt) and audible (heard) response of the vehicle. Shift quality is experienced as various events: Transmission shifts are felt as an upshift at acceleration (1–2), or a downshift maneuver in passing (4–2). Shift engagements of the vehicle are also evaluated, as in Park to Reverse, etc.

Durability / corrosion engineering: Durability and corrosion engineering is the evaluation testing of a vehicle for its useful life. Tests include mileage accumulation, severe driving conditions, and corrosive salt baths.

Package / ergonomics engineering: Package engineering is a discipline that designs/ analyzes the occupant accommodations (seat roominess), ingress/egress to the vehicle, and the driver's field of vision (gauges and windows). The packaging engineer is also responsible for other areas of the vehicle like the engine compartment, and the component to component placement. Ergonomics is the discipline that assesses the occupant's access to the steering wheel, pedals, and other driver/passenger controls.

Climate control: Climate control is the customer's impression of the cabin environment and level of comfort related to the temperature and humidity. From the windshield defrosting to the heating and cooling capacity, all vehicle seating positions are evaluated to a certain level of comfort.

Drivability: Drivability is the vehicle's response to general driving conditions. Cold starts and stalls, RPM dips, idle response, launch hesitations and stumbles, and performance levels.

Cost: The cost of a vehicle program is typically split into the effect on the variable cost of the vehicle, and the up-front tooling and fixed costs associated with developing the vehicle. There are also costs associated with warranty reductions and marketing.

Program timing: To some extent programs are timed with respect to the market, and also to the production schedules of the assembly plants. Any new part in the design must support the development and manufacturing schedule of the model.

Assembly feasibility: It is easy to design a module that is hard to assemble, either resulting in damaged units or poor tolerances. The skilled product development engineer works with the assembly/manufacturing engineers so that the resulting design is easy and cheap to make and assemble, as well as delivering appropriate functionality and appearance.

Quality management: Quality control is an important factor within the production process, as high quality is needed to meet customer requirements and to avoid expensive recall campaigns. The complexity of components involved in the production process requires a combination of different tools and techniques for quality control. Therefore, the International Automotive Task Force (IATF), a group of the world's leading manufacturers and trade organizations, developed the standard ISO/TS 16949. This standard defines the design, development, production, and when relevant, installation and service requirements. Furthermore, it combines the principles of ISO 9001 with aspects of various regional and national automotive standards such as AVSQ (Italy), EAQF (France), VDA6 (Germany) and QS-9000 (USA). In order to further minimize risks related to product failures and liability claims of automotive electric and electronic systems, the quality discipline functional safety according to ISO/IEC 17025 is applied.

Since the 1950s, the comprehensive business approach total quality management, TQM, helps to continuously improve the production process of automotive products and components. Some of the companies who have implemented TQM include Ford Motor Company, Motorola and Toyota Motor Company.

Job Functions

Development Engineer

A development engineer has the responsibility for coordinating delivery of the engineering attributes of a complete automobile (bus, car, truck, van, SUV, motorcycle etc.) as dictated by the automobile manufacturer, governmental regulations, and the customer who buys the product.

Much like the Systems Engineer, the development engineer is concerned with the interactions of all systems in the complete automobile. While there are multiple components and systems in an automobile that have to function as designed, they must also work in harmony with the complete automobile. As an example, the brake system's main function is to provide braking functionality to the automobile. Along with this, it must also provide an acceptable level of: pedal feel (spongy, stiff), brake system "noise" (squeal, shudder, etc.), and interaction with the ABS (anti-lock braking system).

Another aspect of the development engineer's job is a trade-off process required to deliver all of the automobile attributes at a certain acceptable level. An example of this is the trade-off between engine performance and fuel economy. While some customers are looking for maximum power from their engine, the automobile is still required to deliver an acceptable level of fuel economy. From the engine's perspective, these are opposing requirements. Engine performance is looking for maximum displacement (bigger, more power), while fuel economy is looking for a smaller displacement engine (ex: 1.4 L vs. 5.4 L). The engine size however, is not the only contributing factor to fuel economy and automobile performance. Different values come into play.

Other attributes that involve trade-offs include: automobile weight, aerodynamic drag, transmission gearing, emission control devices, handling/roadholding, ride quality, and tires.

The development engineer is also responsible for organizing automobile level testing, validation, and certification. Components and systems are designed and tested individually by the Product Engineer. The final evaluation is to be conducted at the automobile level to evaluate system to system interactions. As an example, the audio system (radio) needs to be evaluated at the automobile level. Interaction with other electronic components can cause interference. Heat dissipation of the system and ergonomic placement of the controls need to be evaluated. Sound quality in all seating positions needs to be provided at acceptable levels.

Manufacturing Engineer

Manufacturing Engineers are responsible for ensuring proper production of the automotive components or complete vehicles. While the development engineers are responsible for the function of the vehicle, manufacturing engineers are responsible for the safe and effective production of the vehicle. This group of engineers consist of Process Engineers, Logistic Coordinators, Tooling Engineers, Robotics Engineers, and Assembly Planners.

In the automotive industry manufacturers are playing a larger role in the development stages of automotive components to ensure that the products are easy to manufacture. Design for Manufacturability in the automotive world is crucial to make certain whichever design is developed in the Research and Development Stage of automotive design. Once the design is established, the manufacturing engineers take over. They design the machinery and tooling necessary to build the automotive components or vehicle and establish the methods of how to mass-produce the product. It is the manufacturing engineers job to increase the efficiency of the automotive plant and to implement lean manufacturing techniques such as Six Sigma and Kaizen.

Other Automotive Engineering Roles

Other automotive engineers include those listed below:

- Aerodynamics engineers will often give guidance to the styling studio so that the shapes they design are aerodynamic, as well as attractive.
- Body engineers will also let the studio know if it is feasible to make the panels for their designs.
- Change control engineers make sure that all of the design and manufacturing changes that occur are organized, managed and implemented.
- Acoustics engineers are specific types of development engineers who do sound and aerodynamic testing to prevent loud cabin noises while the vehicle is on the road.

The Modern Automotive Product Engineering Process

Studies indicate that a substantial part of the modern vehicle's value comes from intelligent systems, and that these represent most of the current automotive innovation. To facilitate this, the modern automotive engineering process has to handle an increased use of mechatronics. Configuration and performance optimization, system integration, control, component, subsystem and system-level validation of the intelligent systems must become an intrinsic part of the standard vehicle engineering process, just as this is the case for the structural, vibro-acoustic and kinematic design. This requires a vehicle development process that is typically highly simulation-driven.

The V-approach

One way to effectively deal with the inherent multi-physics and the control systems development that is involved when including intelligent systems, is to adopt the V-Model approach to systems development, as has been widely used in the automotive industry for twenty years or more. In this V-approach, system-level requirements are propagated down the V via subsystems to component design, and the system performance is validated at increasing integration levels. Engineering of mechatronic systems requires the application of two interconnected "V-cycles": one focusing on the multi-physics system engineering (like the mechanical and electrical components of an electrically powered steering system, including sensors and actuators); and the other focuses on the controls engineering, the control logic, the software and realization of the control hardware and embedded software.

Predictive Engineering Analytics

An alternative approach is called predictive engineering analytics, and takes the V-approach to the next level. It lets design continue after product delivery. That is important for development of built-in predictive functionality and for creating vehicles that can be optimized while being in use, even based on real use data. This approach is based on the creation of a Digital Twin, a replica of the real product that remains in-sync. Manufacturers try to achieve this by implementing a set of development tactics and tools. Critical is a strong alignment of 1D systems simulation, 3D CAE an physical testing to reach more realism in the simulation process. This is combined with intelligent reporting and data analytics for better insight in the vehicle use. By supporting this with a strong data management structure that spans the entire product lifecycle, they bridge the gap between design, manufacturing and product use.

Vehicle Engineering

Vehicle engineering encompasses the fields of automotive engineering, aerospace engineering, rolling stock and marine engineering.

Automotive Engineering

Automotive engineering is the design, manufacture and operation of motorcycles, automobiles and trucks and their respective engineering subsystems. Automotive engineering is one of the most exciting professions you can choose. From the global concerns of sustainable mobility, and teaching cars to drive themselves, to working out how we'll get around on the surface of Mars, automotive engineering is all about the future. The work of an automobile down into three categories:

Design: Designing new products and improving existing ones

Research and Development: Finding solutions to engineering problems

Production: Planning and designing new production processes

Aerospace Engineering

Aerospace engineering is the branch of engineering that deals with the design, development, testing, and production of aircraft and related systems (aeronautical engineering) and of spacecraft, missiles, rocket-propulsion systems, and other equipment operating beyond the earth's atmosphere(astronautical engineering) Further concerned with the science of force and physics that are particular only to performance in Earth's atmosphere and the expanse of space.

Aeronautics

Definition

Aeronautics is the study of the science of flight. Aeronautics is the method of designing an airplane or other flying machine. There are four basic areas that aeronautical engineers must understand in order to be able to design planes. To design a plane, engineers must understand all of these elements.

Design Process

- Aerodynamics: is the study of how air flows around the airplane. By studying the way air flows around the plane the engineers can define the shape of the plane. The wings, the tail, and the main body or fuselage of the plane all affect the way the air will move around the plane.

- Propulsion: is the study of how to design an engine that will provide the thrust that is needed for a plane to take off and fly through the air. The engine provides the power for the airplane. The study of propulsion is what leads the engineers to determine the right kind of engine and the right amount of power that a plane will need.

- Materials and Structures: is the study of what materials are to be used on the plane and in the engine and how those materials make the plane strong enough to fly effectively. The choice of materials that are used to make the fuselage wings, tail and engine will affect the strength and stability of the plane. Many airplane materials are now made out of composites, materials that are stronger than most metals and are lightweight.

- Stability and Control: is the study of how to control the speed, direction, altitude and other conditions that affect how a plane flies. The engineers design the controls that are needed in order to fly and instruments are provided for the pilot in the cockpit of the plane. The pilot uses these instruments to control the stability of the plane during flight.

Astronautics

Astronautics is the design and development of spacecraft with an emphasis on spacecraft systems, the design of ground control systems for spacecraft, and the design of orbital mechanics for spacecraft missions.

Rolling Stock

Rolling stock comprises all the vehicles that move on a railway. It usually includes both powered and unpowered vehicles, for example locomotives, railroad cars, coaches, and wagons.

Naval Architecture

Naval architecture also known as Naval engineering is an engineering discipline dealing with the design, construction, maintenance and operation of marine vessels and structures.

2

An Integrated Study of Automotive Designing

The development of the physical appearance and functionality of road vehicles is known as automotive design. Some of the important disciplines related to this subject are safety engineering, vehicle dynamics, ergonomics engineering, etc. This chapter will provide an integrated understanding of automotive designing.

Automotive Design

Designers at work in 1961. Standing by the scale model's left front fender is Richard Teague, an automobile designer at American Motors Corporation (AMC)

Automotive design is the profession involved in the development of the appearance, and to some extent the ergonomics, of motor vehicles or more specifically road vehicles.

This most commonly refers to automobiles but also refers to motorcycles, trucks, buses, coaches, and vans. The functional design and development of a modern motor vehicle is typically done by a large team from many different disciplines included within automotive engineering. Automotive design in this context is primarily concerned with developing the visual appearance or aesthetics of the vehicle, though it is also involved in the creation of the product concept. Automotive design is practiced by designers who usually have an art background and a degree in industrial design or transportation design.

Design Elements

The task of the design team is usually split into three main aspects: exterior design, interior design, and color and trim design. Graphic design is also an aspect of automotive design; this is generally shared amongst the design team as the lead designer sees fit. Design focuses not only on the isolated outer shape of automobile parts, but concentrates on the combination of form and function, starting from the vehicle package.

A Bertone Birusa concept car on display at an International Car Show.
In the Background are some concept sketches

Draft of OScar design proposal

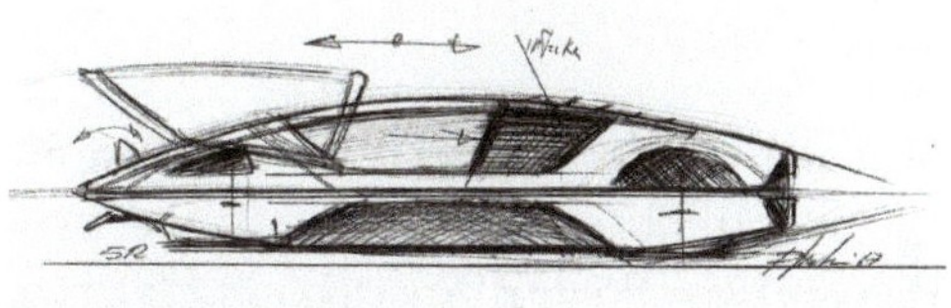

A futuristic original sketch for Ferrari Modulo 512-S by Paul Martin in 1967. There are already many features of the final concept, including the reduced height, wheels coved for low drag and the characteristic entry system.

The aesthetic value will need to correspond to ergonomic functionality and utility features as well. In particular, vehicular electronic components and parts will give more challenges to automotive designers who are required to update on the latest information and knowledge associated with emerging vehicular gadgetry, particularly dashtop mobile devices, like GPS navigation, satellite radio, HD radio, mobile TV, MP3 players, video playback, and smartphone interfaces. Though not all the new vehicular gadgets are to be designated as factory standard items, some of them may be integral to determining the future course of any specific vehicular models.

Exterior Design

The stylist responsible for the design of the exterior of the vehicle develops the proportions, shape, and surfaces of the vehicle. Exterior design is first done by a series of digital or manual drawings. Progressively, drawings that are more detailed are executed and approved by appropriate layers of management. Clay (industrial plasticine) and or digital models are developed from, and along with the drawings. The data from these models are then used to create a full sized mock-up of the final design (body in white). With three- and five-axis CNC milling machines, the clay model is first designed in a computer program and then "carved" using the machine and large amounts of clay. Even in times of high-class 3d software and virtual models on power walls, the clay model is still the most important tool to evaluate the design of a car and therefore used throughout the industry.

Interior Design

The stylist responsible for the design of the vehicle interior develops the proportions, shape, placement, and surfaces for the instrument panel, seats, door trim panels, headliner, pillar trims, etc. Here the emphasis is on ergonomics and the comfort of the passengers. The procedure here is the same as with exterior design (sketch, digital model and clay model).

Color and Trim Design

The color and trim (or color and materials) designer is responsible for the research, design, and development of all interior and exterior colors and materials used on a vehicle. These include paints, plastics, fabric designs, leather, grains, carpet, headliner, wood trim, and so on. Color, contrast, texture, and pattern must be carefully combined to give the vehicle a unique interior environment experience. Designers work closely with the exterior and interior designers.

Designers draw inspiration from other design disciplines such as: industrial design, fashion, home furnishing, architecture, and sometimes product design. Specific research is done into global trends to design for projects two to three model years in the future. Trend boards are created from this research in order to keep track of design influences as they relate to the automotive industry. The designer then uses this information to develop themes and concepts that are then further refined and tested on the vehicle models.

Graphic Design

The design team also develops graphics for items such as: badges, decals, dials, switches, kick or tread strips, liveries.

Computer-aided Styling and Class-A Development

The sketches and rendering are transformed into 3D Digital surface modelling and rendering for real-time evaluation with Math data in initial stages. During the development process succeeding phases will require the 3D model fully developed to meet the aesthetic requirements of a designer and well as all engineering and manufacturing requirements. The fully developed CAS digital model will be re-developed for manufacturing meeting the Class-A surface standards that involves both technical as well as aesthetics. This data will be further developed by Product Engineering team. These modelers usually have a background in Industrial design or sometimes tooling engineering in case of some Class-A modelers. Autodesk Alias and ICEM Surf are the two most widely used software tools for Class-A development.

Development Process

A 1986 Ford Taurus clay model on display in Ford Detroit Styling Studio

Styling Development Cycle

Several manufacturers have slightly varied development cycles for designing an Automobile, but in practice these are the following:

- Design and User Research
- Concept Development sketching
- Class A surfaces
- Clay modeling
- Interior Buck Model
- Vehicle ergonomics
- Class A surfaces

- Colour and Trim
- Vehicle Graphics

The design process occurs concurrently with other product Engineers who will be engineering the styling data for meeting performance, manufacturing and safety regulations. From mid-phase, back and forth interactions between the designers and product engineers culminates into a finished product be manufacturing ready.

Apart from this the Engineering team parallelly works in the following areas. Product Engineering (Body In White Sheetmetal Design and Plastic engineering), NVH Development team, Prototype development, Powertrain engineering, Physical Vehicle validation, Tool and Die development and Manufacturing process design.

Development Team

The styling team for a specific model consists of a chief designer and an Exterior as well as interior designer. In some cases all three roles are done by one designer. Several junior designers are involved in the development process as well who make specific contributions all overseen by the chief designer. Apart from this the Color and trim designer works closely with other designers. The Clay model team and Digital model team works closely with the styling team all located within the studio. Apart from this there would be studio head, studio managers and prototype engineers who would work across all teams in the studio. The total team size for developing a full sedan usually ranges from 25 to 40 members and the development time lasts for more than 24 months till signed-off for tooling and production. Thereafter a smaller team would be working until vehicle launch.

Components

Integration of an automobile involves fitting together separate parts to form a monocoque body or units and mounting these onto a frame, the chassis.

An automobile chassis basically comprises the following:

1. The body shell, which forms the skeleton of the vehicle.
2. The engine, is the power unit of the vehicle; which in the past has been in large part, the internal combustion engine.
3. Transmission system, which aids in transferring the drive from the engine to the wheels. Its main components are the clutch, gearbox, final drive, and differential.
4. Suspension system, which is used to connect the wheels to the body or chassis frame.
5. Steering

6. Brakes
7. Electrical equipment

The chassis is complete in itself as a road vehicle. It can drive and control itself just as in case of a complete car and therefore, in many motor works, the chassis is usually tested on the road before the complete body of the vehicle is attached as the chassis alone can behave as the propulsion means.

History

U.S.

An early example of Sloan's work, 1939 LaSalle

In the United States, automotive design reached a turning point in the 1920's when the American national automobile market began reaching saturation. To maintain unit sales, General Motors head Alfred P. Sloan Jr. suggested annual model-year design changes to convince car owners that they needed to buy a new replacement each year, an idea borrowed from the bicycle industry (though Sloan usually gets the credit, or blame). Critics called his strategy planned obsolescence. Sloan preferred the term "dynamic obsolescence". This strategy had far-reaching effects on the auto business, the field of product design, and eventually the American economy.

An early example of the Forward look design, 1956 Plymouth Fury

A design that was imitated by other automakers, Jeep Cherokee (XJ)

The smaller automakers could not maintain the pace and expense of yearly re-styling. Henry Ford did not like the model-year change because he clung to an engineer's notions of simplicity, economics of scale, and design integrity. GM surpassed Ford's sales in 1931, and became the dominant company in the industry thereafter. The frequent design changes also made it necessary to use a body-on-frame rather than the lighter but less adaptable monocoque design used by most European automakers.

Some U.S. designers include Gordon Buehrig, who was responsible for the Auburn 851, as well as the Cord 810 and 812 (hence also the Hupmobile Skylark and the Graham Hollywood).

In the 1930s, Chrysler's innovations with aerodynamics helped launch the Chrysler Airflow in 1934, which was revolutionary and radical compared to the contemporary vehicles. However, inadequate consumer acceptance of the advanced appearance of the cars forced a re-design of succeeding models of the Airflow. This marketing experience made the entire industry take note of the high risks involved in incorporating major design advancements into their production cars.

A major influence on American auto styling and marketing was Harley Earl, who brought the tailfin and other aeronautical design references to auto design starting with the rear fenders of the 1948 Cadillac. Another notable designer was Chrysler group's designer Virgil Exner, who developed the Forward look design in the mid-1950s. Exner is also credited with using wind tunnel testing to justify incorporating tailfins, thus moving the company away from boxy-looking cars into more aerodynamic and futuristic designs. Raymond Loewy was responsible for a number of Studebaker vehicles, including the Starlight (including the bullet-nose).

Starting in the 1960s, Richard A. Teague, who spent most of his career with American Motors Corporation (AMC), originated the concept of using interchangeable body panels so as to create a wide array of different vehicles using the same stampings, starting with the AMC Cavalier. Teague was responsible for unique automotive designs such as the two-seat AMC AMX muscle car, the subcompact Gremlin, the Pacer, and Matador coupe, as well as the original and market segment-creating, Jeep Cherokee (introduced in 1983, and built to 2001 in the U.S. and through 2005 in China).

Additionally during the 1960s, Ford's first generation Ford Mustang and Thunderbird marked another era leading into new market segments from Detroit. The Ford Mustang achieved record sales in its first year of production and established the pony car segment.

Personal injury litigation has had an effect on the design and appearance of the car in the 20th century.

Europe

Until World War I, most of the automakers were concerned with mechanical reliability rather than its external appearance. Later, luxury and aesthetics became a demand and also an effective marketing tool. Designs from each nation with its own strong cultural identity, reflected in their exterior and interior designs. World War II slowed the progress, but after the early-1950s, Italian designers set the trend and remained the driving force until the early part of the 1980s.

France

In France notable designs came from Bugatti and Avions Voisin. Of the mass selling cars Citroën, launched their vehicles with innovative designs and engineering and mostly aided by the styling of Flaminio Bertoni as evident from Citroën DS. After World War II, with the disappearance of the coach building industry, French automakers (except Citroën) followed British and other popular trends until they gained financial stability. During the 1980s, manufactures like Renault cultivated their own strong design identities with designers like Patrick Le Quement. Peugeot, which was dependent on Pininfarina since early post-war period, later established its own brand identity from the 1980s onwards. Its other company, Citroën, still retains its distinctive French innovations for its designs.

Citroën DS

Great Britain

1981 Ford Sierra with "jelly-mould" or "aero look" (low CD) styling was advanced for its time

Great Britain was Europe's leading manufacturer of automobiles until the late-1960s. During that era there were more British-based automakers than in the rest of Europe combined. The British automobile industry catered to all segments ranging from compact, budget, sports, utility, and luxury-type cars. Car design in Britain was markedly different from other European designs largely because British designers were not influenced by other European art or design movements, as well as the British clay modelers used a different sweep set.

British cars until World War II were sold in most of the British colonies. Innovations in vehicle packaging and chassis engineering combined with global familiarity with British designs meant vehicles were acceptable to public tastes at that time. British skilled resources like panel beaters, die machinists, and clay modelers were also available also partly due their involvement with motorsport industry.

Still during the 1960s, British manufacturers sought professional help from the Italians, such as Giovanni Michelotti, Ercole Spada, and Pininfarina. Notable British contributions to automobile designs were Morris Mini by Alec Issigonis, several Jaguar Cars by Sir William Lyons and Malcolm Sayer, Aston Martin DB Series, and several cars from Triumph and MG. Ford Europe based in Great Britain, is notable for the Ford Sierra line, a creation of Uwe Bahnsen, Robert Lutz, and Patrick le Quément. Other British designers were William Towns for Aston Martin cars and David Bache for Land Rover and Range Rover vehicles.

Germany

Germany is often considered the birthplace of industrial design with Bauhaus School of Design. However, the Nazi regime closed down the design school. Ferdinand Porsche and his family played a significant role in German design. Mercedes Benz passenger cars were also in luxury segment and played more importance to aesthetics. After the 1980s, German design evolved into a distinctive Teutonic style often to complement their high engineered cars suited to Autobahns. The early German design clues of present-day owes some part to Italian designers like Giovanni Michelotti, Ercole Spada, Bruno Sacco, and Giorgetto Giugiaro. During the mid- and late-20th century one of the most influential coach builder/designer in Germany was Karmann.

The 1972 BMW 2002 by Giovanni Michelotti

German designs started gaining popularity after the 1980s, notable after the formation of Audi. Volkswagen, which was dependent on Marcello Gandini, Giorgetto Giugiaro, and Karmann, later formed the contemporary design language along with Audi. BMW entered the automobile design with a sporty-looking everyday sedans using Giovanni Michelotti. These models were later enhanced by Ercole Spada into the 1980s, and Klaus Luthe until the mid-1990s. The American born designer Chris Bangle was hired by BMW in the late-1990s to re-define the brand. Bangle incorporated new single press technology for compound curves to add controversial styling elements to his designs.

The Porsche family contribution were instrumental in the evolution of Porsche cars, while the Italian designer Bruno Sacco helped create various Mercedes Models from the 1960s to the 1990s.

Italy

Ferrari Testarossa from Pininfarina Studios by Leonardo Fioravanti

In Italy, companies like Fiat and Alfa Romeo played a major role in car design. Many coach builders were dependent on these two major manufacturers. Italian manufacturers had a large presence in motorsports leading to several sport car manufacturers like Ferrari, Lancia, Lamborghini, Maserati, etc. During the late-1950s, Italian automobile designs gained global popularity coinciding with the modern fashion and architecture at that time around the world. Various design and technical schools in Turin turned out designers in large scale. By the late-1960s almost all Italian coach builders transformed into design studios catering to automakers around the world. The trend continued in the 1990s, when the Japanese and Korean manufacturers sourced designs from these styling studios. One example is Pininfarina.

Italian designers whose designs services were sought globally are Giovanni Michelotti, Ercole Spada, Bruno Sacco, Marcello Gandini, and Giorgetto Giugiaro.

Scandinavia

Sweden had Volvo and Saab as domestic automakers, and the nation's northern location required that cars needed to withstand Nordic climate conditions. The Scandinavian design elements are known for their minimalism and simplicity. One of the early original Scandinavian designs was the Saab 92001 by Sixten Sason and Gunnar Ljungström.

Ursaab, an early Saab concept illustrating an advanced headlamp treatment

Czechoslovakia

The 1934 Czechoslovakian Tatra T77 is the first serial-produced aerodynamically designed automobile designed by Hans Ledwinka and Paul Jaray

Prior to World War II and until the early-1990s, Czechoslovakia had strong presence in the automotive industry with manufacturers like Skoda, Jawa, Tatra, CZ, Praga, and Zetor. Czech automobiles were generally known for their originality in mechanical simplicity and designs were remarkably Bohemian as evident from Tatra cars and Jawa motorcycles. During the Communist regime, design started falling back and ultimately the domestic automakers ended up as subsidiaries of EU-based companies.

Automotive Hemming

Hemming is a technology used in the automotive industry to join inner and outer closure panels together (hoods, doors, tailgates, etc.). It is the process of bending/folding the flange of the outer panel over the inner one. The accuracy of the operation affects significantly the appearance of the car's outer surfaces and is therefore a critical factor in the final quality of a finished vehicle.

Hemming Processes

Press Hemming

Hemming presses are widely used in automotive manufacturing for hemming of sheet-metal body components. The process uses traditional hydraulically operated 'stamping presses' to hem closure parts and being the last forming process in stamping, it determines the external quality of automotive outer parts such as doors, hood and trunk-lid.

Hemming Press Features and Benefits

- Die storage systems
- Fully automatic die-changing systems with parameters that change with each die
- Pressing capacity typically 60 to 180 tons
- Large panel size capacity
- More than one part can be produced on same line
- Electric hemmers do not need hydraulic oil, are quiet and eco-friendly

Hemming Press Limitations

- Restricted to flat un-complicated panel profiles
- High Cost

Table Top Hemming

Tabletop hemming machines are utilised for the manufacture of medium to high production volumes, with the ability to achieve cycle times as low as 15 seconds.

Table Top Features and Benefits

- Optimum panel quality guaranteed through the hemming principle of the closed ring.

Table Top Limitations

- Dedicated to one panel
- Relatively high cost
- Cost lower than Press Hem, but higher than Robot roller hem

Robot (Roller Hemming)

Robot hemming is utilized for the manufacture of Low to medium production volumes. It uses a standard Industrial Robot integrated with a roller hemming head to provide a flexible solution to closure manufacture. The flange of the outer panel is bent over the inner panel in progressive steps by means of the roller hemming head. The process allows the advantage of using the robot controlled hemming head to hem several different components in a single cell. Minor changes and modifications to panel hemming conditions can also be accommodated allowing a quick and cost-effective reaction.

The robot can also be used for other tasks for example, by equipping it with a tool changing system. This will allow it to also operate dispensing equipment for adhesives and sealants, or a gripper which will enable it to carry out panel manipulation within the assembly cell.

Robot Hemming Features and Benefits

- The ideal solution for all volume production demands, via multiple robots and split operations
- Overall quality better than press or tabletop with MBE hemming technologies implemented
- Flexibility:
  - Robot can hem various panel shapes that press and tabletop can't achieve due to flange attack angles and physical accessibility
  - Can perform panel shape changes via quick program changes, while saving the original programs for future recall
  - Can perform other tasks by changing the hemming tool with another tool such as a dispensing nozzle
- Low cost, simple and quiet in operation when used with standard industrial robot
- Reduced mechanical effort for tryout

Automobile Drag Coefficient

The drag coefficient is a common measure in automotive design as it pertains to aerodynamics. Drag is a force that acts parallel and in the same direction as the airflow. The drag coefficient of an automobile impacts the way the automobile passes through the surrounding air. When automobile companies design a new vehicle they take into consideration the automobile drag coefficient in addition to the other performance characteristics. Aerodynamic drag increases with the square of speed; therefore it becomes critically important at higher speeds. Reducing the drag coefficient in an automobile improves the performance of the vehicle as it pertains to speed and fuel efficiency. There are many different ways to reduce the drag of a vehicle. A common way to measure the drag of the vehicle is through the drag area.

Tatra 77 maquette by Paul Jaray, 1933. This vehicle was the first serial-produced truly aerodynamically designed automobile.

Reducing Drag

The reduction of drag in road vehicles has led to increases in the top speed of the vehicle and the vehicle's fuel efficiency, as well as many other performance characteristics, such as handling and acceleration. The two main factors that impact drag are the frontal area of the vehicle and the drag coefficient. The drag coefficient is a unit-less value that denotes how much an object resists movement through a fluid such as water or air. A potential complication of altering a vehicle's aerodynamics is that it may cause the vehicle to get too much lift. Lift is an aerodynamic force that flows perpendicular to the airflow around the body of the vehicle. Too much lift can cause the vehicle to lose road traction which can be very unsafe. Lowering the drag coefficient comes from streamlining the exterior body of the vehicle. Streamlining the body requires assumptions about the surrounding airspeed and characteristic use of the vehicle.

For high speed applications near or above the speed of sound, a Sears-Haack body, is an idealized shape that minimizes wave drag, which is the drag associated with supersonic shock waves. This shape essentially consists of an elongated tube with pointed ends.

Deletion

The deletion of parts on a vehicle is an easy way for designers and vehicle owners to reduce parasitic and frontal drag of the vehicle with little cost and effort. Deletion can be as simple as removing an aftermarket part, or part that has been installed on the vehicle after production, or having to modify and remove an OEM part, meaning any part of the vehicle that was originally manufactured on the vehicle. Most production sports cars and high efficiency vehicles come standard with many of these deletions in order to be competitive in the automotive and race market, while others choose to keep these drag-increasing aspects of the vehicle for their visual aspects, or to fit the typical uses of their customer base.

Roof Rack

A roof rack is a common trait on many SUV and station wagon vehicles. While roof racks are very useful in carrying extra storage on a vehicle, they also increase the frontal area of the vehicle and increase the drag coefficient. This is because as the air flows over the top of the vehicle, following the smooth lines of the hood and windshield, then collides with the roof rack and causes turbulence. The removal of this part has led to increases in fuel efficiency in several studies.

Mud Flaps

Mudflaps are now rarely specified as standard on production cars as they interfere with the clean airflow around the vehicle. For larger vehicles such as trucks, mud flaps are still important for their control of spray, and in 2010 a new version of the mud flap was introduced that has been shown to create significantly less aerodynamic drag than standard mud flaps.

Rear Spoiler

A rear spoiler usually comes standard in most sports vehicles and resembles the shape of a raised wing in the rear of the vehicle. The main purpose of a rear spoiler in a vehicle's design is to reduce lift, thereby increasing stability at higher speeds. In order to achieve the lowest possible drag, air must flow around the streamlined body of the vehicle without coming into contact with any areas of possible turbulence. A rear spoiler design that stands off the rear deck lid will increase downforce, reducing lift at high speeds while incurring a drag penalty. Flat spoilers, possibly angled slightly downward may reduce turbulence and thereby reduce the coefficient of drag. Some cars now feature automatically adjustable rear spoilers, so at lower speed the effect on drag is reduced when the benefits of reduced lift are not required.

Side Mirrors

Side mirrors both increase the frontal area of the vehicle and increase the coefficient of drag since they protrude from the side of the vehicle. In order to decrease the impact that side mirrors have on the drag of the vehicle the side mirrors can be replaced with smaller mirrors or mirrors with a different shape. Several concept cars of the 2010s are replacing mirrors with tiny cameras but this option is not common for production cars because most countries require side mirrors.

Radio Antenna

While they do not have the biggest impact on the drag coefficient due to their small size, radio antennas commonly found protruding from the front of the vehicle can be relocated and changed in design to rid the car of this added drag. The most common replacement for the standard car antenna is the shark fin antenna found in most high efficiency vehicles.

Windshield Wipers

The effect that windshield wipers have on a vehicles airflow varies between vehicles; however, they are often omitted from race vehicles and high efficiency concepts in order to maintain the smallest possible coefficient of drag. A much more common option is to replace the windshield wipers with lower profile wipers, or to only remove the windshield wiper on the passenger side of the vehicle, and even to fabricate a deflector to deflect the air up and over the wipers.

Another alternative is to equip the vehicle with a single wiper placed in the centre of the windshield, allowing it to cover both sides of the windshield. This mitigates the amount of drag by decreasing the frontal area of the blade. While such application may be useful for racing, for most road vehicles this would produce minimal improvement in overall drag reduction.

Fabrication

The application of new parts and concepts onto the vehicle design are easier to include when in the design stage of a vehicle, rather than in aftermarket (automotive) parts, however, the fabrication of these parts assists in the streamlining of the vehicle and can help greatly reduce the drag of the vehicle. Most vehicles with very low drag coefficients, such as race cars and high efficiency concept cars, apply these ideas to their design.

Wheel Covers

When air flows around the wheel wells it gets disturbed by the rims of the vehicles, and forms an area of turbulence around the wheel. In order for the air to flow smoother around the wheel well smooth wheel covers are often applied. Smooth wheel covers are hub caps with no holes in them for air to pass through. This design reduces drag, however, it may cause the brakes to heat up quicker because the covers prevent airflow around the brake system. This is why this modification is more commonly seen with high efficiency vehicles, rather than sports cars or racing vehicles.

Partial Grille Block

The front grille of a vehicle is used to direct air directly into the engine compartment. In a streamlined design the air flows around the vehicle rather than through; however, the grille of a vehicle redirects airflow from around the vehicle to through the vehicle, which then increases the drag. In order to reduce this impact a grille block is often used. A grille block covers up a portion, or the entirety of the front grille of a vehicle. In most high efficiency models or vehicles with low drag coefficients there will be a very small grille already built into the design, therefore a grille block is unneeded. The grille in most production vehicles is built generally to have as much air flowing into the engine compartment in order to keep it from overheating. But most commonly there is too much airflow into the engine compartment, preventing it from warming up in a timely manner, so a grille block is used to increase engine performance and reduce the vehicle's drag.

Under Tray

The underside of a vehicle often traps air in various places and adds turbulence around the vehicle. In most racing vehicles this is eliminated by covering the entire underside of the vehicle in what is called an under tray. This tray prevents any air from becoming trapped under the vehicle and reduces drag.

Fender Skirts

Fender skirts are often made as extensions of the body panels of the vehicles and cover the entire wheel wells. Much like smooth wheel covers this modification reduces the drag of the vehicle by preventing any air from becoming trapped in the wheel well and assists in streamlining the body of the vehicle. Fender skirts are more commonly found on the rear wheel wells of a vehicle because the tires do not turn and the design is much simpler. This is commonly seen in vehicles such as the first generation Honda Insight. Front fender skirts have the same effect on reducing drag as the rear wheel skirts, but must be further offset from the body in order to compensate for the tire sticking out from the body of the vehicle as turns are made.

Modified Front Bumper

The front bumper is the first part of the vehicle that the air must flow around. Therefore, it plays a crucial role in reducing drag. A front air dam is often used which extends from the very front of the vehicle down to the lowest part of the vehicle. This is done to direct airflow around and over the vehicle rather than allowing air to travel under it. Contoured deflectors, or tire spats, are often made as part of the front bumper in order to direct airflow around the tire without having any increase to the outward flow.

Boattails and Kammbacks

A boattail can greatly reduce a vehicle's total drag. Boattails create a teardrop shape that will give the vehicle a more streamlined profile, reducing the occurrence of drag inducing flow separation. A kammback is a truncated boattail. It is created as an extension of the rear of the vehicle, moving the rear backward at a slight angle toward the bumper of the car. This can reduce drag as well but a boattail would reduce the vehicles drag more. Nonetheless, for practical and style reasons, a kammback is more commonly seen in racing, high efficiency vehicles, and trucking.

Automotive Suspension Design

Automotive suspension design is an aspect of automotive engineering, concerned with designing the suspension for cars and trucks. Suspension design for other vehicles is similar, though the process may not be as well established.

The process entails:

- Selecting appropriate vehicle level targets
- Selecting a system architecture
- Choosing the location of the 'hard points', or theoretical centres of each ball joint or bushing
- Selecting the rates of the bushings
- Analysing the loads in the suspension
- Designing the spring rates
- Designing shock absorber characteristics
- Designing the structure of each component so that it is strong, stiff, light, and cheap
- Analysing the vehicle dynamics of the resulting design

Since the 1990s the use of multibody simulation and finite element software has made this series of tasks more straightforward.

Vehicle Level Targets

A partial list would include:

- Maximum steady state lateral acceleration (in understeer mode)
- Roll stiffness (degrees per g of lateral acceleration)
- Ride frequencies
- Lateral load transfer percentage distribution front to rear

- Roll moment distribution front to rear
- Ride heights at various states of load
- Understeer gradient
- Turning circle
- Ackermann
- Jounce travel
- Rebound travel

Once the overall vehicle targets have been identified they can be used to set targets for the two suspensions. For instance, the overall understeer target can be broken down into contributions from each end using a Bundorf analysis.

System Architecture

Typically a vehicle designer is operating within a set of constraints. The suspension architecture selected for each end of the vehicle will have to obey those constraints. For both ends of the car this would include the type of spring, location of the spring, and location of the shock absorbers.

For the front suspension the following need to be considered:

- The type of suspension (MacPherson strut or double wishbone suspension)
- Type of steering actuator (rack and pinion or recirculating ball)
- Location of the steering actuator in front of, or behind, the wheel centre

For the rear suspension there are many more possible suspension types, in practice.

Hardpoints

The hardpoints control the static settings and the kinematics of the suspension.

The static settings are:

- Toe
- Camber
- Caster
- Roll center height at design load
- Mechanical (or caster) trail
- Anti-dive and anti-squat

- Kingpin Inclination
- Scrub radius
- Spring and shock absorber motion ratios

The kinematics describe how important characteristics change as the suspension moves, typically in roll or steer. They include:

- Bump Steer
- Roll Steer
- Tractive Force Steer
- Brake Force Steer
- Camber gain in roll
- Caster gain in roll
- Roll centre height gain
- Ackermann change with steering angle
- Track gain in roll

The analysis for these parameters can be done graphically, or by CAD, or by the use of kinematics software.

Compliance Analysis

The compliance of the bushings, the body, and other parts modify the behaviour of the suspension. In general it is difficult to improve the kinematics of a suspension using the bushings, but one example where it does work is the toe control bush used in Twist-beam rear suspensions. More generally, modern cars suspensions include a Noise, vibration, and harshness (NVH) bush. This is designed as the main path for the vibrations and forces that cause road noise and impact noise, and is supposed to be tunable without affecting the kinematics too much.

In racing cars, bushings tend to be made of harder materials such as brass or delrin.

In general physical terms, the mass and mechanical hysteresis (damping effect) of solid parts should be accounted for in a dynamic analysis, as well as their elasticity.

Loads

Once the basic geometry is established the loads in each suspension part can be estimated. This can be as simple as deciding what a likely maximum load case is at the contact patch, and then drawing a Free body diagram of each part to work out the forces, or as complex as simulating the behaviour of the suspension over a rough road, and calculating the loads caused. Often loads that have been measured on a similar suspension are used instead - this is the most reliable method.

Detailed Design of Arms

The loads and geometry are then used to design the arms and spindle. Inevitably some problems will be found in the course of this that force compromises to be made with the basic geometry of the suspension.

Class A Surfaces

Class A surfaces is a term used in automotive design to describe a set of freeform surfaces of high efficiency and quality. Although, strictly, it is nothing more than saying the surfaces have curvature and tangency alignment - to ideal aesthetical reflection quality, many people interpret class A surfaces to have G2 (or even G3) curvature continuity to one another.

Class A surfacing is done using computer-aided industrial design applications. Class A surface modellers are also called "digital sculptors" in the industry. Industrial Designers develop their design styling through the A surface, the physical surface the end user can feel, touch, see etc.

The Class A surfacing is also termed as "Digital Sculpting" with numerous software tools now available in the automotive industry. The notable software tools are Alias by Autodesk, Icemsurf, Catia by Dassault systems etc.

A common method of working is to start with a prototype model and produce smooth mathematical Class A surfaces to describe the product's outer body. From this the production of tools and inspection of finished parts can be carried out. Class A surfacing complements the prototype modelling stage by reducing time and increasing control over design iterations.

Class A surfaces can be defined as any surface, that has styling intent, that is either seen, touched or both.

In an automotive application that means everything in the interior cabin and all of the exterior excluding the under body. This can also include beauty covers in the engine compartment as well as trunk panels and carpeting.

In the product design realm, Class A surfacing can be applied to such things like a housing for a hand drill, a plastic laundry detergent bottle, toys or furniture.

Aerospace has some styling considerations like bezels for air vents and lights.

Clay Modelling

Class A surfacing / digital sculpting is similar to clay modelling with the added advantage of computing power to change or incorporate design changes in existing/new design. Moreover, the revisions of clay modelling and refinement iteration are carried out in digital version. The scanned data of a selected clay model will be taken as a "Point cloud data" input, and Class A designers work on this Point cloud data to generate preliminary surfaces and further refine them to Class A surfaces.

Cab Over

Cab-over, also known as Cab Over Engine (COE), cab forward (U.S.), or forward control (UK), is a body style of truck, bus, or van that has a vertical front, "flat face" or a semi-hood, with the cab of the truck sitting above (or forward of) the front axle. This contrasts with a conventional truck where the engine is mounted in front of the driver.

This truck configuration is currently common among European, Japanese and Chinese truck manufacturers, because the laws governing overall vehicle lengths are strict and the body style allows longer trailers (in the case of a tractor-trailer truck) or a longer cargo area (in the case of a "straight" truck, a truck with a single rigid frame supporting all components and the load) for the same overall length.

Although popular among United States heavy truckers and trucking companies during the 1970s because of strict length laws in many states, when those length laws were repealed, most heavy-truck makers moved to other body styles. It is, however, still very popular in the light- and medium-duty truck segment. However American companies Paccar and Freightliner still manufacture traditional cab over engine designs for the Australian and South African markets where length restrictions still make them advantageous.

In Australia both American (cab over axle) and European/Japanese/Chinese (cab forward of axle) types, as well as the conventional type are common. Cab over engine types dominate urban and light duty use, with conventional trucks predominating in remote and off road areas. Both types are common for highway use.

History

The first truck in the United States was built by Autocar in 1899 using a format then called "engine-under-the-seat" and was available with optional 5 or 8 horsepower motors. Although early Autocar trucks were not exactly "cab-over", since the truck did not have a cab, per se, they were the fore-runners to COEs. The Sternberg company of Wisconsin produced cab-over trucks as early as 1907, though by 1914 only their seven-ton model was a cab-over. They reintroduced the cab-over layout in 1933 with their "Camel Back" model, which allowed the cab to be tilted to access the engine.

Vintage COE: 1942 Chevrolet.

Preserved International Harvester Metro Van in Portland in 2012

The introduction of the first modern cab-over layout in the United States is credited to industrial designer Viktor Schreckengost, who, with engineer Ray Spiller, designed a cab-over truck for the White Motor Company in 1932. Schreckengost was later joined by other designers such as Raymond Loewy who designed the Metro series of vans and trucks for International Harvester. The bodies for these vehicles were initially produced by the Metropolitan Body Company (MBC). The company produced a wide variety of truck and commercial bodies for several vehicle manufacturers such as Chevrolet, Ford, Dodge Bros., and International Harvester until 1948 when they were purchased by the latter. MBC was instrumental in the development of COE route delivery bodies in the 1930s.

The laws of the time limited overall truck length to 42 feet (12.8 m) on highways. Setting the cab over the engine and front axle shaved several feet off the length of the tractor, feet which could then be added to the length of the trailer while keeping the dimensions of the entire truck within the permissible limit. Schreckengost patented the design in 1934.

Autocar reintroduced the engine-under-the-seat format with their Model U in 1933, which became a staple of U.S. roads and the U.S. military through World War II and well into the 1950s.

White-Freightliner introduced its first tilting cab-over design in 1958, which allowed the entire cab to tilt forward for access to the engine.

Design Aspects

1962 presented Sisu KB-112/117 was the first European serial produced truck with a hydraulically tiltable cabin, enabling an easy access to the engine.

In Class 8 tractors (using the US designation), the cab-over design allows the vehicle's wheelbase to be shorter than in the conventional arrangement, wherein the engine is placed in front of the cab, covered by a horizontal or sloping hood that opens to allow engine access. Its shorter wheelbase allows cab-over semi trucks to have a shorter overall length, thereby allowing for longer trailers to be used. For light- and medium-duty solid- or rigid-axle trucks, the cab-over design requires less length for the cab and engine, in a given wheelbase, and therefore allows a greater length for the truck body or load area.

In both class 8 tractors and light- and medium-duty vocational trucks, the cab-over-engine design gives the COE model an advantage in maneuverability over a conventional model. And since COEs are generally lighter than conventionals, they can often haul heavier loads, given equal gross vehicle weight rating (GVWRs) and gross combination weight rating (GCWRs). Despite the COE designs' being smaller in general, over-the-road tractors can still be fully equipped with single or bunk beds. Also, lack of a hood gives better visibility to the driver and a tighter turning radius, and significantly reduces the forward blind spots.

One critique is that the shorter wheelbase in the COE semi-trucks gives a rougher ride than those with conventional cabs, as the driver's seat is above the front axle; and that the cabs tend to be noisier because the engine is directly below, though this may be alleviated by noise-damping insulation.

Because of their flat front design, early COE semi-trucks had significantly worse aerodynamics than conventional tractors. Modern cab-over designs, in both semi-trucks and light- and medium-duty models, have improved aerodynamics significantly over early models, but often still have higher drag coefficients than their modern conventional-design counterparts. This works against fuel economy, and offsets some of the improvement in fuel consumption garnered from the lighter weight of the cab-over truck when running less than fully loaded.

Although the tilting cab gives comparatively unobstructed access to the engine, its deployment causes unsecured items in the cab and sleeper (if equipped) to fall onto the windshield or under the instrument panel. Vehicles without a tilting cab will usually be equipped with removable floor panels through which mechanics can access and service the engine. Unfortunately, some components might require servicing or replacement that either will not fit through the hatch or remain inaccessible. This necessitates removal of the engine by lowering it out from under the vehicle, a time-consuming and expensive procedure.

Safety

In early cab-over models, the lack of a safe crumple zone in the front made them far more dangerous in the event of a crash. Modern designs employ stiffening beams in the doors to help prevent collapse during a front-end collision, steering columns are typically collapsible, all models include seat belts, and some models are equipped with airbags. All of these features improve occupant safety during a collision.

In military use, the COE design increases the chance of crew injury or death in the event of running over a landmine. This is because the tire that detonates the mine is directly below/beside the driver.

Mercedes-Benz Actros semi-trailer truck

Tatra T815 8x8 cab-over-in-front autocrane

Land Rover 101 Forward Control with radio-vehicle body

Car Classification

Governments and private organizations have developed car classification schemes that are used for innumerable purposes including regulation, description and categorization, among others.

Classification Methods

Vehicles can be categorized in numerous ways. For example, by means of the body style and the "level of commonality in vehicle construction as defined by number of doors and roof treatment (e.g., sedan, convertible, fastback, hatchback) and number of seats" that require seat belts to meet safety regulations.

Regulatory agencies may also establish a vehicle classification system for determining a tax amount. In the United Kingdom, a vehicle is taxed according to the vehicle's construction, engine, weight, type of fuel and emissions, as well as the purpose for which it is used. Other jurisdictions may determine vehicle tax based upon environmental principles, such as the user pays principle. In another example, certain cities in the United States in the 1920s chose to exempt electric-powered vehicles because officials believed those vehicles did not cause "substantial wear upon the pavements."

Another standard for road vehicles of all types that is used internationally (except for Australia, India, and the U.S.), is ISO 3833-1977.

In an example from private enterprise, many car rental companies use the ACRISS Car Classification Code to describe the size, type and equipment of vehicles to ensure that rental agents can match customer needs to available vehicles, regardless of distance between the agent and the rental company or the languages spoken by either party. In the United States, since 2010 the Insurance Institute for Highway Safety uses a scheme it has developed that takes into account a combination of both vehicle shadow (length times width) and weight.

Highway Loss Data Institute classification	Definition
Regular Two Door	Two door sedans and hatchbacks
Regular Four Door	Four door sedans and hatchbacks
Station Wagons	Four doors, a rear hatch and four pillars
Minivans	Vans with sliding rear doors
Sports	Two seaters and cars with significant high performance features
Luxury	Relatively expensive cars that are not classified as sports (price in USD to curb weight in pounds more than 9.0 in 2010) (small cars over $27,000, midsize cars over $31,500, large cars over $36,000, etc.)

Highway Loss Data Institute 'Guide to car size groups' (IIHS includes minivans in this group)					
	Shadow (square footage of exterior length x width)				
Curb Weight	**70 to 80 sq ft (6.503–7.432 m²)**	**81 to 90 sq ft (7.525–8.361 m²)**	**91 to 100 sq ft (8.454–9.290 m²)**	**101 to 110 sq ft (9.383–10.22 m²)**	**>110 sq ft (10.22 m²)**
2,001 to 2,500 lb (908–1,134 kg)	Mini	Small	Small	Small	Midsize
2,501 to 3,000 lb (1,134–1,361 kg)	Small	Small	Midsize	Midsize	Midsize
3,001 to 3,500 lb (1,361–1,588 kg)	Small	Midsize	Midsize	Large	Large
3,501 to 4,000 lb (1,588–1,814 kg)	Small	Midsize	Large	Large	Very Large
>4,000 lb (1,814 kg)	Midsize	Midsize	Large	Very Large	Very Large

IIHS|HLDI Guide to SUV size groups	
	curb weight
Small	3,001 to 3,750 lb (1,361–1,701 kg)
Midsize	3,751 to 4,750 lb (1,701–2,155 kg)
Large	4,751 to 5,750 lb (2,155–2,608 kg)
Very large	5,750 lb (2,608 kg) or shadow > 115 sq ft (10.68 m²)

The United States National Highway Traffic Safety Administration (NHTSA) separates vehicles into classes by the curb weight of the vehicle with standard equipment including the maximum capacity of fuel, oil, coolant, and air conditioning, if so equipped.

NHTSA classification	Code	Curb weight
Passenger cars: mini	PC/Mi	1,500 to 1,999 lb (680–907 kg)
Passenger cars: light	PC/L	2,000 to 2,499 lb (907–1,134 kg)
Passenger cars: compact	PC/C	2,500 to 2,999 lb (1,134–1,360 kg)
Passenger cars: medium	PC/Me	3,000 to 3,499 lb (1,361–1,587 kg)
Passenger cars: heavy	PC/H	3,500 lb (1,588 kg) and over
Sport utility vehicles	SUV	–
Pickup trucks	PU	–
Vans	VAN	–

The United States Federal Highway Administration has developed a classification scheme used for automatically calculating road use tolls. There are two broad categories depending on whether the vehicle carries passengers or commodities. Vehicles that carry commodities are further subdivided by number of axles and number of units, including both power and trailer units.

The United States Environmental Protection Agency (US EPA) has developed a classification scheme used to compare fuel economy among similar vehicles. Passenger vehicles are classified based on a vehicle's total interior passenger and cargo volumes. Trucks are classified based upon their gross vehicle weight rating (GVWR). Heavy duty vehicles are not included within the EPA scheme.

EPA car class	Total passenger and cargo volume (cu. ft.)
Two-seaters	Any (designed to seat only two adults)
Minicompact	Less than 85 cu ft (2,407 l)
Subcompact	85 to 99 cu ft (2,407–2,803 l)
Compact	100 to 109 cu ft (2,832–3,087 l)
Mid-size	110 to 119 cu ft (3,115–3,370 l)
Large	120 cu ft (3,398 l) or more
Small station wagons	Less than 130 cu ft (3,681 l)
Mid-size station wagons	130 to 159 cu ft (3,681–4,502 l)
Large station wagons	160 cu ft (4,531 l) or more

A similar set of classes is used by the Canadian EPA. The Canadian National Collision Database (NCDB) system defines "passenger car" as a unique class, but also identifies two other categories involving passenger vehicles—the "passenger van" and "light utility vehicle"—and these categories are inconsistently handled across the country with the boundaries between the vehicles increasingly blurred.

In Australia, the Federal Chamber of Automotive Industries publishes its own classifications.

Size and Usage-based Vehicle Classification Systems Worldwide

This is a summary table listing several different methods of vehicle classification.

<table>
<tr><th colspan="8">Vehicle Classification</th></tr>
<tr><th colspan="3">Not Well Defined / Vernacular</th><th colspan="4">Defined by Law or Regulation</th><th rowspan="2">Examples</th></tr>
<tr><th>Market Segment (American English)</th><th>Market Segment (British English)</th><th>Market Segment (Australian English)</th><th>US EPA Size Class</th><th>Euro NCAP Structural Category</th><th>Euro NCAP Class (1997 - 2009)</th><th>Euro Market Segment</th></tr>
<tr><td>Microcar</td><td>Microcar, Bubble car</td><td>N/A</td><td>N/A</td><td>—</td><td>Quadricycle</td><td rowspan="2">A-segment mini cars</td><td>Bond Bug, Isetta, Mega City, Renault Twizy, REVAi/G-Wiz</td></tr>
<tr><td rowspan="2">Subcompact car Economy Car</td><td>City car</td><td>Microcar</td><td>Minicompact</td><td rowspan="12">Passenger car</td><td rowspan="2">Supermini</td><td>Citroën C1, Fiat 500, Hyundai Eon, Mitsubishi i-MiEV, Renault Twingo, Smart Fortwo</td></tr>
<tr><td>Supermini</td><td>Light Car</td><td>Subcompact</td><td>B-segment small cars</td><td>Ford Fiesta, Kia Rio, Opel Corsa, Peugeot 208, Volkswagen Polo</td></tr>
<tr><td>Compact car</td><td>Small family car</td><td>Small Car</td><td>Compact</td><td>Small family car</td><td>C-segment medium cars</td><td>Honda Civic, Mazda3, Peugeot 308, Renault Mégane, Toyota Corolla</td></tr>
<tr><td>Mid-size car</td><td>Large family car</td><td>Medium Car</td><td>Mid-size</td><td rowspan="2">Large family car</td><td rowspan="2">D-segment large cars</td><td>Chevrolet Malibu, Ford Fusion, Peugeot 508, Subaru Legacy, Volkswagen Passat</td></tr>
<tr><td>Entry-level luxury car</td><td>Compact executive car</td><td>Medium Car above $60,000</td><td>N/A</td><td>Acura ILX, Alfa Romeo Giulia, Audi A4, Lexus ES, Mercedes-Benz C-Class</td></tr>
<tr><td>Full-size car</td><td rowspan="2">Executive car</td><td>Large Car</td><td>Large</td><td rowspan="2">Executive</td><td rowspan="2">E-segment executive cars</td><td>Chevrolet Impala, Ford Taurus, Mazda Xedos 9, Hyundai Grandeur, Holden Commodore, first and second generation Škoda Superb, Toyota Avalon</td></tr>
<tr><td>Mid-size luxury car</td><td>Large Car above $70,000</td><td>N/A</td><td>Audi A6, Cadillac CTS, Chrysler 300, Tesla Model S, Acura TLX</td></tr>
<tr><td>Full-size luxury car</td><td>Luxury car</td><td>Upper Large Car above $100,000</td><td>N/A</td><td>—</td><td>F-segment luxury cars</td><td>BMW 7 Series, Lincoln Town Car, Mercedes-Benz S-Class, Porsche Panamera, Audi A8, Maserati Quattroporte</td></tr>
<tr><td>Grand tourer</td><td>Grand tourer</td><td rowspan="4">Sports Car</td><td>N/A</td><td>—</td><td rowspan="4">S-segment sports coupés</td><td>Aston Martin DB9, Bentley Continental GT, Ferrari FF, Jaguar XK, Maserati GranTurismo</td></tr>
<tr><td>Supercar</td><td>Supercar</td><td>N/A</td><td>—</td><td>Bugatti Veyron, LaFerrari, Lamborghini Aventador, Pagani Zonda, Porsche 918 Spyder</td></tr>
<tr><td>Convertible</td><td>Convertible</td><td>N/A</td><td>—</td><td>BMW 6 Series, Chevrolet Camaro, Mercedes CLK, Volvo C70, Volkswagen Eos</td></tr>
<tr><td>Roadster</td><td>Roadster</td><td>Two-seater</td><td>Roadster sports</td><td>BMW Z4, Lotus Elise, Mazda MX-5, Porsche Boxster, Mercedes-Benz SLK</td></tr>
</table>

<table>
<tr><td>—</td><td>Mini MPV</td><td>N/A</td><td rowspan="3">Minivan</td><td rowspan="5">MPV</td><td rowspan="2">Small MPV</td><td rowspan="5">M-segment multi purpose cars</td><td>Citroen C3 Picasso, Ford B-Max, Opel Meriva, Fiat 500L</td></tr>
<tr><td>MPV</td><td>Compact MPV</td><td rowspan="2">People Mover</td><td>Chevrolet Orlando, Ford C-Max, Opel Zafira, Renault Scenic, Volkswagen Touran</td></tr>
<tr><td>Minivan</td><td>Large MPV</td><td>Large MPV</td><td>Chrysler Town and Country, Kia Carnival, Citroën C4 Grand Picasso, Renault Espace, Toyota Sienna</td></tr>
<tr><td>Cargo van</td><td>Van</td><td>Van</td><td>Cargo van</td><td>—</td><td>Chevrolet Express 1500 Cargo, Fiat Ducato/Ram ProMaster, Ford Transit, Renault Master, Volkswagen Transporter</td></tr>
<tr><td>Passenger van</td><td>Minibus</td><td>People Mover</td><td>Passenger van</td><td>—</td><td>Chevrolet Express 1500 Passenger, Ford E350 Wagon, Mercedes-Benz Viano</td></tr>
<tr><td>Mini SUV</td><td>Mini 4x4</td><td>Small SUV</td><td rowspan="2">Small Sport Utility Vehicle</td><td rowspan="4">Off-roader</td><td rowspan="2">Small Off-Road 4x4</td><td rowspan="4">J-segment sport utility cars (including off-road vehicles)</td><td>Daihatsu Terios, Ford Ecosport, Jeep Renegade, Peugeot 2008, Suzuki Jimny</td></tr>
<tr><td>Compact SUV</td><td>Compact SUV</td><td>Medium SUV</td><td>Audi Q5, Chevrolet Equinox, Ford Escape, Honda CR-V, Jeep Cherokee, Alfa Romeo Stelvio</td></tr>
<tr><td>Mid-size SUV</td><td rowspan="2">Large 4x4</td><td>Large SUV</td><td rowspan="2">Standard Sport Utility Vehicle</td><td rowspan="2">Large Off-Road 4x4</td><td>Audi Q7, Ford Expedition, Hyundai Santa Fe, Jeep Grand Cherokee, Volkswagen Touareg, Maserati Levante</td></tr>
<tr><td>Full-size SUV</td><td>Upper Large SUV</td><td>Range Rover, Cadillac Escalade, Toyota Land Cruiser</td></tr>
<tr><td>Mini pickup truck</td><td rowspan="4">Pick-up</td><td rowspan="4">Pick-up</td><td rowspan="2">Small Pickup Truck</td><td rowspan="4">Pickup</td><td rowspan="4">Pick-up</td><td rowspan="4">—</td><td>Chevrolet Montana, Fiat Strada, Renault Duster Oroch, Volkswagen Saveiro</td></tr>
<tr><td>Mid-size pickup truck</td><td>Ford Ranger, Chevrolet Colorado, Mitsubishi Triton/L200, Nissan Navara, Toyota Hilux</td></tr>
<tr><td>Full-size pickup truck</td><td rowspan="2">Standard pickup truck</td><td>Dodge Ram, Ford F-150, GMC Sierra, Nissan Titan, Toyota Tundra</td></tr>
<tr><td>Heavy Duty pickup truck</td><td>Chevrolet Silverado HD, Ram Heavy Duty, Ford Super Duty</td></tr>
<tr><td>Special purpose vehicle</td><td>—</td><td>Limousine</td><td>Special purpose vehicle</td><td>—</td><td>—</td><td>—</td><td>Lincoln MKT Livery</td></tr>
</table>

Microcar

Straddling the boundary between car and motorbike, these vehicles have engines under 1.0 litre, typically seat only two passengers, and are sometimes unorthodox in construction. Some microcars are three-wheelers, while the majority have four wheels. Microcars were popular in post-war Europe, where their appearance led them to be called "Bubble cars". More recent microcars are often electric powered.

Microcar Abaca

Examples of microcars:

- Isetta
- Smart Fortwo
- Tata Nano

Hatchbacks

Ultracompact Car

In 2012, Japan's Transport and Tourism Ministry allowed local government to use ultracompact cars as transport for residents and tourists in their limiting areas. The size of ultracompact cars will be less than minicars, but have engine greater than 50cc displacement and able to transport 1 or 2 persons. Ultracompact cars cannot use minicars standard, because of strict safety standards for minicars. The regulation about running capacity and safety performance of ultracompact cars will be published in early autumn. Today, there are cars smaller than ultracompact cars, called category-1 motorized vehicles which it has 50cc displacement or less and only one seat for the driver.

City Car

Citroën C1

A city car is a small automobile intended for use in urban areas. Unlike microcars, a city car's greater speed, capacity and (in perception at least) occupant protection are safer in mixed traffic environments and weather conditions. While city cars can reach highway speeds, that is not their intended use. In Japan, city cars are called kei cars. Kei cars have to meet strict size and engine requirements: engines have a maximum displacement of 660 cc and the car's length must be under 3400 mm.

Examples of kei cars:

- Daihatsu Move
- Honda Life
- Suzuki Cervo

Examples of city cars:

- Fiat Panda
- Maruti 800
- Mini (Original 1959 model)

Other small cars:

- Carver One
- Citroën Type C
- Smith Flyer

Supermini/Subcompact Car

Renault Clio IV

This class is known as supermini in the UK, subcompact in North America. Superminis have three, four or five doors, and even as an estate shape. They are designed to seat four passengers comfortably. Current supermini hatchbacks are approximately 3900 mm long, while saloons and estate cars are around 4200 mm long. Currently (2013) sedan variants are generally not available in Europe and are marketed at a lower price than hatchback models in North America.

In Europe, the first superminis were the Fiat 500 of 1957 and the Austin Mini of 1959. Superminis can be premium cars, such as the Citroën DS3, named 2010 Car of the Year by Top Gear Magazine. Superminis are some of the best selling vehicles in Europe with 25% of the market shares (2013). In 2007, the Peugeot 207 has been the most sold car in Europe, whereas the best seller is almost systematically a car from the *compact* segment.

In Australia, the motoring press tends to distinguish between a light car such as the Daihatsu Charade or early models of the Holden Barina, and slightly larger models such as the Ford Fiesta which is considered to be a small car. As the general size of vehicles in this class has gradually increased, the category of light car has almost disappeared.

Examples of superminis/subcompact cars:

- Opel Corsa
- Peugeot 208
- Volkswagen Polo

This category is equivalent to the EuroNCAP class "Superminis".

Family Car

Small Family Car/Compact Car

Volkswagen Golf

Small family/compact cars refer to the hatchbacks and shortest saloons and estate cars with similar size. They are approximately 4,250 mm (167 in) long in case of hatchbacks and 4,500 mm (177 in) in the case of saloons and estate cars. Compact cars have room for five adults and usually have engines between 1.4 and 2.2 litres, but some have engines of up to 2.5 litres. Some early "muscle" compacts had optional V8 engines of up to 6.6 liters. These are the most popular vehicles in most developed countries.

Examples of hatchback small family cars/compact cars:

- Peugeot 308
- Toyota Auris
- Renault Megane

This category is equivalent to the EuroNCAP class "Small Family Cars". In Australia, this class is generally referred to as being small-medium sized cars.

Large Family / Mid-size

Citroën DS5

Traditionally, mid-size cars are sedans, but recently cars such as the Citroën DS5, which is a large hatchback family car, have introduced other body styles.

Saloons / Sedans

Large Family / Mid-size

Toyota Camry

A class described as "large family" in Europe and "mid-size" in the USA, these cars have room for five adults and a large trunk (boot). Engines are more powerful than small family/compact cars and six-cylinder engines are more common than in smaller cars. Car sizes vary from region to region; in Europe, large family cars are rarely over 4,700 mm (15.4 ft) long, while in North America, Middle East and Australasia they may be well over 4,800 mm (15.7 ft).

Examples of large family cars/mid-size cars:

- Chevrolet Malibu
- Ford Mondeo
- Kia Optima

This category is equivalent to the EuroNCAP class "Large Family Cars". These are known in Australia as Medium sized cars.

Full Size / Large

Holden Commodore

This term is used most in North America, Middle East and Australia where it refers to the largest affordable sedans on the market. Full-size cars may be well over 4,900 mm (16.1 ft) long.

Examples of full-size cars:

- Dodge Charger
- Ford Falcon
- Toyota Avalon

Crossover SUV

Mitsubishi Outlander

Crossover SUVs are derived from an automobile platform using a monocoque construction with light off-road capability and lower ground clearance than SUVs. They may be styled similar to conventional "off-roaders", or may be look similar to an estate car or station wagon.

Examples of crossover SUVs:

- Chevrolet Equinox
- Nissan Qashqai
- Tata Aria

Minivans / MPVs

Renault Espace, one of the first true minivans

Also known as "people carriers", this class of cars resembles tall estate cars. Larger MPVs may have seating for up to eight passengers. (Beyond that size, similar vehicles tend to be derived from vans and in Europe are called minibuses.)

Being taller than a family car improves visibility for the driver (while reducing visibility for other road users) and may help access for the elderly or disabled. They also offer more seats and increased load capacity than hatchbacks or estate cars.

Examples of mini MPVs:

- Citroën C3 Picasso
- Ford B-Max
- Nissan Note

Examples of compact MPVs:

- Opel Zafira Tourer
- Peugeot 5008
- Renault Scénic

Both categories are equivalent to the EuroNCAP class "Small MPVs".

Examples of large MPVs / minivans:

- Dodge Caravan

- Ford S-Max
- Mazda5

This category is equivalent to the EuroNCAP class “MPVs”.

Luxury Vehicle

Compact Executive

Lexus IS

These are luxurious equivalents to mid-size and compact cars. Rear seat room and trunk space are smaller than executive cars simply because of their smaller overall size.

Examples of compact premium cars/entry-level luxury cars:

- Audi A4
- BMW 3 Series
- Buick Regal

This category is equivalent to the EuroNCAP class “Large Family Cars”.

Executive/Mid-luxury

BMW 5 Series

These are luxurious equivalents to full-size cars. This also refers to the largest hatchbacks within the similar length in this class, such as the Porsche Panamera.

Examples of executive cars/mid-luxury cars:

- Peugeot 607
- Jaguar XF
- MG Magnette

This category is equivalent to the EuroNCAP class "Executive Cars".

Full-size Luxury / Grand Saloon

Mercedes-Benz S-Class

Also known as full-size luxury cars, grand saloons, or premium large cars, while "Oberklasse" is used in Germany. Typically a four-door saloon (sedan). These are the most powerful saloons, with six, eight and twelve-cylinder engines and have more equipment than smaller models. Vehicles in this category include some of the models from the flagship lines of luxury car brands, such as Cadillac CT6, Lincoln Town Car and Maserati Quattroporte.

Examples of grand saloons:

- Audi A8
- Lexus LS
- BMW 7-Series

This category is equivalent to the EuroNCAP class "Executive Cars".

Estate Cars / Station Wagons

A station wagon (also known as an estate or estate car) is an automobile with a body style variant of a sedan/saloon with its roof extended rearward over a shared passenger/cargo volume with access at the back via a third or fifth door (the liftgate or tailgate), instead of a trunk lid. The body style transforms a standard three-box design into a two-box design—to include an A, B, and C-pillar, as well as a D-pillar. Station wagons can flexibly reconfigure their interior volume via fold-down rear seats to prioritize either passenger or cargo volume.

Examples of estates/station wagons:

- Hyundai i40 Tourer
- Jaguar XF Sportbrake
- Mercedes-Benz CLS Shooting Brake

Sports Cars

Hot Hatch

Peugeot 205 GTI crowned "The Greatest Ever Hot Hatch"

A hot hatch is a high-performance hatchback, based on standard superminis or small family cars with improved performance, handling and styling. Hot hatches are very popular in Europe, where hatchbacks are by far the most common body style for this size of car. In North America, sport compacts are usually sold as saloons or coupés rather than hatchbacks.

Examples of hot hatches:

- Volkswagen Golf GTi
- Peugeot 205 GTi
- Fiat 500 Abarth

Sports Saloon / Sports Sedan

Pontiac G8 GT

These are high-performance versions of saloons. Sometimes originally homologated for production based motorsports (touring cars or rally cars) and like regular saloons, seats four or five people.

Examples of sports saloons/sedans:

- BMW M5
- Mazdaspeed6/Mazda 6 MPS
- Dodge Charger

Examples of sport compact saloons/sedans:

- Dodge SRT-4
- Lotus Cortina
- Mitsubishi EVO

Sports Car

Jaguar E-Type

The term "sports car" does not appear to have a clear definition. It is commonly used to describe vehicles which prioritise acceleration and handling; however, some people claim it is also defined as a vehicle with two seats.

A Sports car (sportscar or sport car) is a small, usually two-seat, two-door automobile designed for spirited performance and nimble handling. Sports cars may be spartan or luxurious but high maneuverability and minimum weight are requisite.

Examples of sports cars:

- Chevrolet Corvette
- Mazda MX-5
- Porsche 911

Grand Tourer

Maserati GranTurismo

Larger, more powerful and heavier than sports cars, these vehicles typically have a FR layout and seating for four passengers (2+2). These are more expensive than sports cars but not as expensive as supercars. Grand Tourers encompass both luxury and high-performance. Some grand tourers are hand-built.

Examples of grand tourers:

- Aston Martin V8
- Lexus SC300/400
- Ferrari 612 Scaglietti

Supercar

Lamborghini Countach

Supercar is a term generally used for ultra-high-end exotic cars, whose performance is superior to that of its contemporaries. The proper application of the term is subjective and disputed, especially among enthusiasts.

Examples of supercars:

- McLaren P1
- Koenigsegg Agera R
- Bugatti Veyron 16.4

Muscle Car

1970 The Machine

The muscle car term generally refers to rear wheel drive mid-size cars with powerful V8 engines, manufactured in the U.S. Some definitions limit it to two-door vehicles; however, others include four-door body style versions. Although opinions vary, it is generally accepted that classic muscle cars were produced in the late 1960s and early 1970s. Muscle cars were also produced in Australia and other nations.

Examples of American muscle cars from the 1960s and 1970s:

- Ford Torino
- Plymouth Road Runner
- Pontiac GTO

Examples of Australian muscle cars:

- Ford Falcon
- Holden Monaro
- Valiant Charger

Pony Car

1966 Ford Mustang

The pony car is a class of American automobile launched and inspired by the Ford Mustang in 1964. It describes an affordable, compact, highly styled car with a sporty or performance-oriented image.

Examples of pony cars:

- AMC Javelin
- Chevrolet Camaro
- Dodge Challenger

Convertible

Full-sized convertible with its fabric covered top folded behind the rear seat

A body design that features a flexibly operating roof for open or enclosed mode driving. Also known as a cabriolet or roadster (if a 2-seater). Historically, convertibles used folding roof structures with fabric or other flexible materials. Some designs have roofs made of metal or other stiff materials that retract into the body.

Examples of cabriolets:

- Mazda MX-5
- Honda S2000
- Volvo C70

Off-roaders

Off-road vehicles, or "off-roaders" are sometimes referred to as "four-wheel drives", "four by fours", or 4x4s — this can happen colloquially in cases where certain models or even an entire range does not possess four-wheel drive.

Sport Utility Vehicle

Jeep Commander

Ford Kuga

Sport utility vehicles are off-road vehicles with four-wheel drive and true off-road capability. They most often feature high ground clearance and an upright, boxy body design. Sport Utilities are typically defined by a body on frame construction which offers more off-road capability but reduced on-road ride comfort and handling compared to a cross-over or car based utility vehicle.

Examples of compact SUVs:

- Land Rover Freelander

- Jeep Patriot
- Toyota FJ Cruiser

This category is equivalent to the EuroNCAP class "Small Off-Roaders".

Examples of SUVs:

- Land Rover Discovery
- Mitsubishi Pajero
- Mahindra Scorpio

This category is equivalent to the EuroNCAP class "Large Off-Roaders".

Commercial Vehicle

Van

American conversion van

In some countries, the term "van" can refer to a small *panel van* based on a passenger car design (often the estate model / station wagon); it also refers to light trucks, which themselves are sometimes based on SUVs or MPVs. (But note that those retaining seats and windows, while being larger and more utilitarian than MPVs, may be called "mini-buses".) The term is also used in the term "camper van" (or just "camper") – equivalent to a North American recreational vehicle (RV).

In the United States, the term "van" refers to vehicles that, like European minibuses, are even larger than large MPVs and are rarely seen being driven for domestic purposes – except for "conversion vans". These possess extremely large interior space and are often more intended for hauling cargo than people. Most vans use body-on-frame construction and are thus suitable for extensive modification and coachwork, known as conversion. Conversion vans are often quite luxurious, boasting comfortable seats, soft rides, built-in support for electronics such as television sets, and other amenities. The more elaborate conversion vans straddle the line between cars and recreational vehicles.

Examples of North American "vans":

- Dodge Ram Van
- Ford E-Series
- GMC Savana

Examples of European "vans":

- Ford Transit
- Volkswagen Transporter
- Mercedes-Benz Sprinter

Examples of Japanese "vans"

- Toyota Hiace
- Nissan NV

Other Car Classification Terms

Bakkie

A generic South African term for light pickup truck.

Baquet

Refers to cars made in the early 1900s in Europe. Baquet means bath tub. These cars had two rows of raised seats similar to horse-drawn carriages. Baquets usually did not have front doors, a top, or windshield.

Buggy

A Buggy is an automobile with wheels that project beyond the vehicle body.

Cabrio coach

Normally a two-door body design with special form of car roof, where a retractable textile cover amounts to a large sunroof.

Coupé

A 2-door, 2- or 4-seat car with a fixed roof. Its doors are often longer than those of an equivalent sedan and the rear passenger area smaller; the roof may also be low. In cases where the rear seats are very small and not intended for regular use it is called a 2+2 (pronounced *"two plus two"*). Originally, a coupé was required to have only one side window per side, but this consideration has not been used for many years.

Coupé utility

A passenger-car derived vehicle with an integral exterior cargo area.

Crossover (or CUV)

A loose marketing term to describe a vehicle that blends features of a SUV with features of a car — especially forgoing the body on frame construction of the SUV in favor of the car's unibody or monocoque construction. Crossovers usually borrow drivetrains and other parts from traditional cars in the same manufacturer's line. Crossovers typically employ an FF layout or an FF-based four-wheel drive layout with a transverse engine, rather than an FR layout or an FR-based 4WD layout with a longitudinal engine as is typically used on traditional truck-based SUVs.

Drop Head Coupe

Generally a European term referring to a 2-door, 4 place automobile with a retractable canvas / cloth top with *both* a padded headliner and rollup windows (as opposed to side curtains).

Estate

British name for a *station wagon.*

Fastback

A design where the roof slopes at a smooth angle to the tail of the car, but the rear window does not open as a separate "door".

Flower Car

In US, similar to ute in Australia, i.e. generic for Chevy El Camino, Ford Ranchero, GMC Sprint/Diablo, etc.

Hatchback

Incorporates a shared passenger and cargo volume, with rearmost accessibility via a rear third or fifth door, typically a top-hinged liftgate—and features such as fold-down rear seats to enable flexibility within the shared passenger/cargo volume. As a two-box design, the body style typically includes A, B and C-pillars, and may include a D-pillar.

Hardtop

Originally a removable solid roof on a *convertible*; later, also a fixed-roof car whose doors have no fixed window frames, which is designed to resemble such a convertible.

Hearse

A converted car (often a station wagon), light truck or minivan usually used to transport the dead. Often longer and heavier than the vehicle on which they are usually based. Can sometimes double up as an ambulance in some countries, such as the United States, especially in rural areas.

Kammback

Originally, a car with a tapered rear that cuts off abruptly.

Landaulet

A limousine with the passenger section covered by a convertible top.

Leisure activity vehicle

A small van, generally related to a supermini, with a second or even a third seat row, and a large, tall boot.

Liftback

A broad marketing term for a hatchback, which incorporates a shared passenger and cargo volume, with rearmost accessibility via a top-hinged liftgate.

Limousine

By definition, a chauffeur-driven car with a (normally glass-windowed) division between the front seats and the rear. In German, the term simply means a sedan.

Microvan

Term for a boxy wagon-type of car that is smaller than a conventional minivan; often without rear sliding door(s). Examples are Citroën Picasso, Renault Scénic, Toyota Yaris Verso or Mercedes-Benz A-Class. In Japan, this term is used for Kei car based vans.

Minibus

Designed to carry fewer people than a full-size bus, generally up to 16 people in multiple rows of seats. Passenger access in normally via a sliding door on one side of the vehicle. One example of a van with a minibus version available is the Ford Transit.

MPV

Multi-purpose vehicle, a large car or small bus designed to be used on and off-road and easily convertible to facilitate loading of goods from facilitating carrying people.

Notchback

A configuration where the third box of a three-box styling configuration is less pronounced — especially where the rear deck (third box) is short or where the rear window is upright.

People carrier or people mover

European name to describe what is usually referred to in North America as a Minivan.

Phaeton

A Phaeton is a style of open car or carriage without proper weather protection for passengers.

Pickup truck (or *pickup*)

A light-duty, open-bed truck.

Pillarless

Usually a prefix to *coupé*, *fastback*, or *hardtop*; completely open at the sides when the windows are down, without a central pillar, e.g. the Sunbeam Rapier fastback coupé.

Ragtop

Originally an open car like a *roadster*, but with a soft top (cloth top) that can be raised or lowered. Unlike a convertible, it had no roll-up side windows. Now often used as slang for a convertible.

Retractable Hardtop

Aka Coupé convertible or Coupé Cabriolet. A type of convertible forgoing a foldable textile roof in favor of a multi-segment rigid roof retracts into the lower bodywork.

Roadster

Originally a two-seat open car with minimal weather protection — without top or side glass — though possibly with optional hard or soft top and side curtains (i.e., without roll-up glass windows). In modern usage, the term means simply a two-seat sports car *convertible*, a variation of *spyder*.

Sedan

A car seating four or more with a fixed roof that is full-height up to the rear window. Known in British English as a saloon. Sedans can have 2 or 4-doors. This is the most common body style. In the U.S., this term has been used to denote a car with fixed window frames, as opposed to the *hardtop* style wherein the sash, if any, winds down with the glass.

Sedan delivery

North American term for a vehicle similar to a *wagon* but without side windows, similar to a panel truck but with two doors (one on each side), and one or two rear doors.

Sport utility vehicle (SUV)

Derivative of a pickup truck or 4-wheel-drive vehicle, but with fully enclosed passenger cabin interior and carlike levels of interior equipment.

Spyder (or Spider)

Similar to a *roadster* but originally with less weather protection.Nowadays it simply means a convertible with two seater only. The name comes from the old carriages with two seats and no roof, whose small central cabin and big wheels at the corners are reminiscent of a spider.

Shooting-brake

Initially a vehicle used to carry shooting parties with their equipment and game; later used to describe custom-built wagons by high-end coachbuilders, subsequently synonymous with *station wagon* or *estate*; and in contemporary usage a three or five-door wagons combining features of a wagon and a coupé.

Station wagon

A variant of a sedan/saloon, (also known as estate or estate car) or with its roof extended rearward over a shared passenger/cargo volume; access at the back via a third or fifth door instead of a trunk lid; flexible configurations to vary passenger or cargo volume; and two or three rows of seating — in a two-box design with a A, B & C-pillar, as well as a D pillar.

T-top

A derivative of the Targa top, called a T-bar roof, this fixed-roof design has two removable panels and retains a central narrow roof section along the front to back axis of the car (*e.g.* Toyota MR2 Mark I.)

Targa top

A semi-convertible style used on some sports cars, featuring a fully removable hard top roof panel which leaves the A and B pillars in place on the car body.

Town car (US)

Essentially the inverse of the landaulet, a historical body style in which the front seats were open and the rear compartment closed, normally with a removable top to cover the front chauffeur's compartment. In Europe the style is also known as Sedanca de Ville, often shortened to Sedanca or de Ville. Note that the modern Lincoln Town Car derives its name, but nothing else, from this style.

Ute

A term used originally in Australia and New Zealand to describe usually two-wheel-drive, traditionally passenger vehicles with a cargo tray in the rear integrated with the passenger body; as opposed to a pickup whose cargo tray is not integrated with the passenger body.

Wagon delivery

North American term (mainly U.S. and Canada). Similar to a *sedan delivery*, with four doors.

Van

In North America "van" refers to a truck-based commercial vehicle of the wagon style, whether used for passenger or commercial use. Usually a van has no windows at the side rear (panel van), although for passenger use, side windows are included. In other parts of the world, 'van' denotes a passenger-based wagon with no rear side windows.

Non-English Terms

Some non-English language terms are familiar from their use on imported vehicles in English-speaking nations even though the terms have not been adopted into English.

Barchetta

Italian term for a roadster with no roof. The name, roughly "small boat", comes from an exclamation when the Ferrari 166MM Touring was shown.

Berlina

Italian term for a sedan.

Berline

French term for a sedan.

Berlinetta

Italian term for a sport coupé.

Break

French term for a station wagon.

Camioneta

Brazilian Portuguese term for a station wagon (specially in the state of Rio de Janeiro). Spanish term also used in Argentina and Uruguay.

Carrinha

Portuguese term for a station wagon. Not used in Brazilian Portuguese.

Espada

Portuguese nickname for a limousine (the same word for *Sword* – long piece of metal). Not used in Brazilian Portuguese.

Furgoneta

Spanish and Polish term for a van, in the latter language almost always used in its diminutive form *furgonetka.*

Furgão

Portuguese alternative term (less used) for a van. Used in Brazilian Portuguese, most often for vans but sometimes for panel van variants of passenger cars.

Kombi

is a German abbreviation of "Kombinationswagen" (Combination Car) and it is German name for station wagon. Since Germany is a major producer of cars for many European countries, the term Kombi in this meaning is also used in Swedish, Czech, Slovak, Polish, Slovenian, Serbian, Bosnian, Croatian, Hungarian, Spanish, Portuguese, Bulgarian. In Afrikaans and in Australia, Kombi is also used to refer to a Volkswagen Microbus. In Brazil, Argentina and Uruguay the word specifically refers to the VW Microbus.

Perua

Brazilian Portuguese term either designating a van (especially as spoken in the city of São Paulo) or a station wagon (in the city of Rio de Janeiro).

Turismo

Spanish term for a sedan. Literally means tourism, used mostly in Latin American countries and Spain.

Ijapa

Yoruba term for a two door car. Literally modeled after a Tortoise animal. Also refers to Volkswagen Beetle.

Sport Utility Vehicle

1995–2001 Ford Explorer built on a light-truck chassis

A sport utility vehicle or suburban utility vehicle (SUV) is a vehicle classified as a light truck, but operated as a family vehicle. SUV's are similar to a large station wagon or estate car, though typically featuring tall interior packaging, high H-point seating, high center of gravity, high ground-clearance and especially four- or all-wheel-drive capability for on- or off-road ability. Some SUVs include the towing capacity of a pickup truck with the passenger-carrying space of a minivan or large sedan.

Popular in the late-1990s and early–mid-2000s, SUVs sales temporarily declined due to high oil prices and a declining economy. The traditional truck-based SUV is gradually being supplanted by the crossover SUV, which uses an automobile platform for lighter weight and better fuel efficiency. By 2010, SUV sales around the world were growing, in spite of high gas prices.

Definitions

According to the Merriam-Webster dictionary, a "sport utility vehicle" is "a rugged automotive vehicle similar to a station wagon but built on a light-truck chassis". The "SUV" term is defined as "a large vehicle that is designed to be used on rough surfaces but that is often used on city roads or highways." The "SUV" acronym "is still used to describe nearly anything with available all-wheel drive and raised ground clearance."

North America

There is no one definition for an SUV. Most government regulations simply have categories for "off-highway vehicles," which in turn are lumped in with pickup trucks and minivans as "light trucks." The auto industry has not settled on one definition.

Nevertheless, four-wheel-drive SUVs are considered light trucks in North America (and two-wheel-drive SUVs up to the 2011 model year) where they were regulated less strictly than passenger cars under two laws in the United States, the Energy Policy and Conservation Act for fuel economy, and the Clean Air Act for emissions. Starting in 2004, the United States Environmental Protection Agency (EPA) began to hold sport utility vehicles to the same tailpipe emissions standards as cars.

Many people question "how can an SUV be called a truck?" Although the original definition of the "light truck" classification included pickups and delivery vans, usually SUVs and minivans are included in this category because these vehicles are designed to "permit greater cargo-carying capacity than passenger carrying volume." Manufacturing, emissions, and safety regulations in the U.S. classify "an SUV is a truck"; however, for local licensing and traffic enforcement, "an SUV may be a truck or a car" because the classification of these vehicles varies from state to state. For industry production statistics, SUVs are counted in the light truck product segment.

Other Markets

The term is not used in all countries, and outside North America the terms "off-road vehicle", "four-wheel drive" or "four-by-four" (abbreviated to "4WD" or "4×4") or simply use of the brand name to describe the vehicle like "Jeep" or "Land Rover" are more common.

In Europe, the term SUV has a similar meaning, but being newer than in the U.S. it only applies to the newer street oriented one, where-as "Jeep", "Land Rover" or 4x4 are used for the off-roader oriented ones. Not all SUVs have four-wheel drive capabilities, and not all four-wheel-drive passenger vehicles are SUVs. Although some SUVs have off-road capabilities, they often play only a secondary role, and SUVs often do not have the ability to switch among two-wheel and four-wheel-drive high gearing and four-wheel-drive low gearing. While automakers tout an SUV's off-road prowess with advertising and naming, the daily use of SUVs is largely on paved roads.

In India, all SUVs are classified in the "Utility Vehicle" category per the Society of Indian Automobile Manufacturers (SIAM) definitions and carry a 27% excise tax. Those that are 4 metres (157 inches) long, have a 1,500 cc (92 cu in) engine or larger, along with 170 mm (6.7 in) of ground clearance, are subject to a 30% excise duty.

Designs

Although designs vary, SUVs have historically been mid-size passenger vehicles with a body-on-frame chassis similar to that found on light trucks. Early SUVs were mostly two-door models, and were available with removable tops. However, consumer demand pushed the SUV market towards four doors, by 2002 all full-size two-door SUVs were gone from the market. Two-door SUVs were mostly carry-over models, and their sales were not viable enough to warrant a redesign at the end of their design cycle. The Jeep Wrangler remained as a compact two-door body style, although it was also joined by a four-door variant starting with the 2007 model year, the Wrangler Unlimited. The number of two-door SUV models increased in the 2010s with the release of the Range Rover Evoque and the Nissan Murano convertible, although both vehicles are unibody.

Most SUVs are designed with an engine compartment, a combined passenger and cargo compartment, and no dedicated trunk such as in a station wagon body. Most mid-size and full-size SUVs have three rows of seats with a cargo area directly behind the last row of seats. Cargo barriers are often fitted to the cargo area to protect the vehicles occupants from injury from unsecured cargo in the event of sudden deceleration or collision.

SUVs are known for high ground clearance, upright, boxy body, and high H-point. This can make them more likely to roll over due to their high center of gravity. Bodies of SUVs have recently become more aerodynamic, but the sheer size and weight keeps their fuel economy poor.

Mini SUV

Honda Vezel (Mini SUV)

A mini SUV (also called subcompact SUV or subcompact crossover) is a class of small sport utility vehicles. The term usually applies to crossovers based on a supermini (B-segment cars in Europe) platform such as the Daihatsu Terios, Chevrolet Trax, Fiat 500X, Honda Vezel, Jeep Renegade, Mini Countryman, Mercedes-Benz GLA-Class, Opel Mokka, Peugeot 2008 and Renault Captur.

Compact SUV

Audi Q3 (Compact SUV)

A compact SUV is a class of smaller SUVs that are commonly built with less cargo and passenger space, and often with smaller engines resulting in better fuel economy, the term is often interchangeable with crossover SUV. Some examples are: Acura RDX, Audi Q3, Audi Q5, BMW X1, BMW X3, Chery Tiggo, Chevrolet Captiva, Ford Escape, Great Wall Haval H5, Honda CR-V, Honda Element, Honda Passport, Isuzu Rodeo, Jeep Compass, Jeep Cherokee, Mahindra Quanto, Mazda CX-5, Mercedes-Benz GLK-Class, Mercedes-Benz GLC-Class, Mitsubishi ASX, Opel Antara, Porsche Macan, Range Rover Evoque, Renault Koleos, Škoda Yeti, Suzuki Vitara, Toyota RAV4, Volkswagen Tiguan and Volvo XC60.

Mid-size SUV

BMW X5 (Mid-Size SUV)

A mid-size SUV is a class of medium-size SUVs whose size typically falls between that of a full-size and a compact SUV. This term is not commonly used outside North America, where fullsize and midsize SUVs are considered similar. Some examples are: Acura MDX, Audi Q7, Buick Rendezvous, BMW X5, BMW X6, Chevrolet Equinox, Chevrolet TrailBlazer, Dodge Journey, Ford Edge, Ford Explorer, Honda Pilot, Hummer H3, Hyundai Santa Fe, Jeep Grand Cherokee, Jeep Liberty, Land Rover Discovery, Mazda CX-7, Mercedes-Benz M-Class, Mitsubishi Challenger, Nissan Murano, Nissan Pathfinder, Nissan Xterra, Pontiac Aztek, Pontiac Torrent, Range Rover Sport, Tata

Safari, Toyota 4Runner, Toyota Highlander, Toyota FJ Cruiser, Volkswagen Touareg, Ssangyong Rexton, Tesla Model X and Volvo XC90.

Full-size SUV

Nissan Patrol (Full Size SUV)

Full-size SUVs have greater cargo and passenger space than midsize SUVs. Full Size SUVs are usually given higher safety ratings than their smaller counterparts. Some examples are: Cadillac Escalade, Chevrolet Tahoe, Dodge Durango, Ford Expedition, GMC Yukon, Hummer H2, Infiniti QX56, Jeep Wagoneer, Land Rover Discovery, Lexus LX, Lincoln Navigator, Mitsubishi Montero, Mercedes-Benz G-Class, Mercedes-Benz GL, Nissan Armada, Nissan Patrol, Toyota Land Cruiser and Toyota Sequoia.

Extended-length SUV

Cadillac Escalade ESV (Extended-Length SUV)

An extended length SUV, also sometimes called a long-wheel based SUV, are vehicles that are similar to a full-size SUV, except that these vehicles have a larger cargo area (around 130 in (3.30 m)) and passenger space that can seat up to 8 or 9 people (with the available third row seating that when folded or removed adds more cargo space). Although these extended length SUVs are mostly sold in North America because of their size and the roads are made and designed differently, they can also be found in other countries, exported to such places like The Philippines and The Middle East. The vehicles are 221 in (5.61 m) to 223 in (5.66 m) in length and can be distinguished by the rear wheel area not touching the rear doors. Currently, the only vehicles built under this segment are the Cadillac Escalade ESV, Chevrolet Suburban, Ford Expedition EL, Ford Excursion, GMC Yukon XL and Lincoln Navigator L.

History

Early SUVs were descendants from commercial and military vehicles such as the World War II Jeep and Land Rover.

The earliest examples of longer-wheelbase wagon-type cars were the Chevrolet Carryall Suburban (1935, RWD only), GAZ-61 (1938, 4×4), Willys Jeep Station Wagon (1948), Pobeda M-72 (GAZ-M20/1955), which Russian references credit as possibly being the first modern SUV (with unitary body rather than body-on-frame), International Harvester Travelall (1953), Land Rover Series II 109 (1958), and the International Harvester Scout 80 (1961). These were followed by the more 'modern' Jeep Wagoneer (1963), International Harvester Scout II (1971), Ford Bronco (1966), Toyota Land Cruiser FJ-55 (1968), the Chevrolet Blazer / GMC Jimmy (1969), and the Land Rover Range Rover (1970). The actual term "sport utility vehicle" did not come into wide popular usage until the late 1980s; many of these vehicles were marketed during their era as station wagons.

According to Robert Casey, the transportation curator at the Henry Ford Museum, the Jeep Cherokee (XJ) was the first true sport utility vehicle in the modern understanding of the term. Developed under the leadership of AMC's François Castaing and marketed to urban families as a substitute for a traditional car (and especially station wagons, which were still fairly popular at the time), the Cherokee had four-wheel drive in a more manageable size (compared to the full-size Wagoneer), as well as a plush interior resembling a station wagon. With the introduction of more luxurious models and a much more powerful 4-liter engine, sales of the Cherokee increased even higher as the price of gasoline fell, and the term "sport utility vehicle" began to be used in the national press for the first time. "The advent and immediate success of AMC/Jeep's compact four-door Cherokee turned the truck industry upside down."

The corporate average fuel economy (CAFE) standard was ratified in the 1970s to regulate the fuel economy of passenger vehicles. Car manufacturers evaded the regulation by selling SUVs as work vehicles. The popularity of SUV increased among urban drivers in the last 25 years, and particularly in the last decade. Consequently, modern SUVs are available with luxury vehicle features, and some crossover models adopt lower ride heights to accommodate on-road driving.

Keith Bradsher explained the rise of the SUV with American Motors' (AMC) lobbying the United States Environmental Protection Agency (EPA) for a waiver of the United States Clean Air Act. The EPA subsequently designated AMC's compact Cherokee as a "light truck", and the company marketed the vehicle to everyday drivers. AMC's effort to affect rulemaking changing the official definition of their new model then led to the SUV boom when other auto makers marketed their own models in response to the Cherokee taking sales from their regular cars.

Jeep Cherokee: SUV trendsetter as designed by AMC

Popularity

SUVs became popular in the United States, Canada, India and Australia in the 1990s and early-2000s. U.S. automakers could enjoy profit margins of $10,000 per SUV, while losing a few hundred dollars on a compact car. For example, the Ford Excursion could net the company $18,000, while they could not break even with the Ford Focus unless the buyer chose options, leading Detroit's big three automakers to focus on SUVs over small cars.

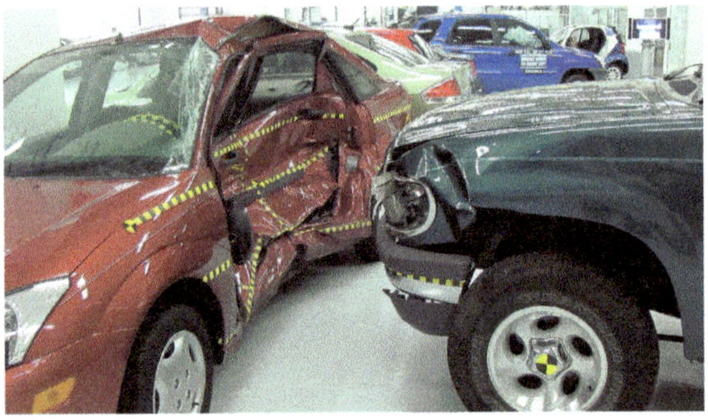

A crash test by the Insurance Institute for Highway Safety shows the damage to a compact Ford Focus struck by a Ford Explorer SUV

The higher cost of union labor in the U.S. and Canada compared to the lower wages of non-union workers at non-U.S. companies like Toyota, made it unprofitable for American auto makers to build small cars in the U.S. For example, the General Motors factory in Arlington, Texas where rear-wheel-drive cars were built, such as the Chevrolet Caprice, Buick Roadmaster, and Cadillac Fleetwood Brougham was converted to truck and SUV production, putting an end to full-size family station wagon and overall terminating production of rear-wheel drive full-size cars. Due to the shift in the Big Three's strategy, many long-running cars like the Ford Taurus, Buick Century and Pontiac Grand Prix fell behind their Japanese competitors in features and image (relying more on fleet sales instead of retail and/or heavy incentive discounts); some were discontinued.

Buyers were drawn to SUVs' large cabins, higher ride height, and perceived safety. Full-size SUVs often offered features such as three-row seating, to effectively replace full-size station wagons and minivans. Wagons were seen as old-fashioned. Additionally, full-size SUVs have greater towing capabilities than conventional cars, and can haul trailers, travel trailers (caravans) and boats. Increased ground clearance is useful in climates with heavy snowfall. The very low oil prices of the 1990s helped keep down running costs. The SUV was one of the most popular choices of vehicle for female drivers in the U.S.

Social scientists have drawn on popular folklore such as urban legends to illustrate how marketers have been able to capitalize on the feelings of strength and security offered by SUVs. Popular tales include narratives where mothers save the family from armed robbery and other incidents by taking the automobile off road, for example.

In Australia, SUV sales were helped by the fact that SUVs had much lower import duty than passenger cars did, so that they cost less than similarly equipped imported sedans. However, this gap was gradually narrowed, and in January 2010 the import duty on cars was lowered to match the 5 percent duty on SUVs.

Sales of SUVs and other light trucks fell in the mid-2000s because of high oil prices and declining economy. In 2008, General Motors announced plans to close four truck and SUV plants, including the Oshawa Truck Assembly. The company cited decreased sales of large vehicles in the wake of rising fuel prices. The business model of focusing on SUVs and light trucks, at the expense of more fuel-efficient compact and midsized cars, is blamed for declining sales and profits among Detroit's Big Three automakers since the mid–late-2000s. The Big Three were slower to adapt than their Japanese rivals in producing small cars to meet growing demand due to inflexible manufacturing facilities, which made it unprofitable to build small cars. However, starting in 2010 SUV and light truck sales have started an upward trend due to lower gas prices and a revival of the North American economy. In 2013, General Motors saw its sales for its large SUVs increased by 74%, making them the largest producer of SUVs in the United States. However, the "small and compact SUVs, when compared with other vehicles in the light truck segment, has made this vehicle segment the third highest selling vehicle segment in the automotive market in 2013." With the redesigned GM and Ford large SUVs being introduced in 2014 (for the 2015 model year), it has seen a slight resurgence among consumers due to better fuel economy and new engines, along with updated and newer features.

Starting in 2015, sales of SUVs started dominating the industry. At the end of 2016, sales of SUVs and light duty trucks had surpassed traditional car sales for the year by over 3 million units. Manufacturers like Hyundai have started reducing their production of traditional cars in favor of SUVs citing reduction in sales and difficulty competing with other manufacturers.

Use in Remote Areas

SUVs are sometimes driven off-road on farms and in remote areas of such places as the Australian Outback, Africa, the Middle East, Alaska, Canada, Iceland, South America, Russia and parts of Asia which have limited paved roads and require a vehicle to have all-terrain handling, increased range, and storage capacity. The scarcity of spare parts and the need to carry out repairs quickly resulted in the popularity of vehicles with the bare minimum of electric and hydraulic systems, such as the basic versions of the Land Rover, Jeep Wrangler, Nissan Patrol and Toyota Land Cruiser. SUVs for urban driving have traditionally been developed from their more rugged all-terrain counterparts. For example, the Hummer H1 was developed from the HMMWV, originally developed for the military of the United States.

As many SUV owners never used the off-road capabilities of their vehicle, newer SUVs (including crossovers) now have lower ground clearance and suspension designed primarily for paved roads.

Some buyers choose SUVs because they have more interior space than sedans of similar sizes. In areas with gravel roads in summer and snow and ice in winter, four-wheel drives offer a safety advantage due to their traction advantages under these conditions.

The sport utility vehicles have also gained popularity in some areas of Mexico, especially in desert areas or in cities where drivers frequently encounter potholes, detours, high water and rough roads. Increasing use is also attributed to the high number of dirt roads outside major population centers, resulting in washboard and mud in the rainy seasons.

Use in Recreation and Motorsport

Some highly modified SUVs, together with their more rugged off-road counterparts, are also used to explore places otherwise unreachable by other vehicles. In Australia, China, Europe, South Africa, South America and the United States at least, 4WD clubs have been formed for this purpose. Modified SUVs also take part in races, including the Paris-Dakar Rally, the Baja racing series, TREC events, King of the Hammers in California and the Australian Outback.

The Trophee Andros ice-racing series is another competition where SUVs participate as well.

Many 4×4 mud racing events and other activities take place throughout the US organized by clubs and associations.

Luxury SUV

Numerous luxury vehicles in the form of SUVs and pickup trucks are being produced. Luxury SUV is principally a marketing term to sell fancier vehicles that may have higher performance, comfort, technology, or brand image. The term lacks both measurability and verifiability, and it is applied to a broad range of SUV sizes and types.

Nevertheless, the marketing category was created in 1966 with Kaiser Jeep's luxurious Super Wagoneer. It was the first SUV to offer a V8 engine, automatic transmission, and luxury car trim and equipment in a serious off-road model. It came with bucket seating, air conditioning, sun roof, and even a vinyl roof. Land Rover followed suit in 1970 by introducing the Range Rover. The trend continued with other competitors adding comfort features to their rudimentary and truck-based models.

The production of luxury models increased in the late-1990s with vehicles such as the Lincoln Navigator and Cadillac Escalade. These luxury SUVs generated higher profit margins than non-luxury SUVs did. For some auto makers, luxury SUVs were the first SUV models they produced. Some of these models are not traditional SUVs based on light truck as they are classified as crossovers.

The luxury SUV class encompasses both smaller 5-passenger SUVs and larger 7-passenger SUVs, with luxury features both inside of the cabin but also in the outside. Buyers looking for a luxury vehicle that offers more cargo capacity than a sedan may prefer a luxury SUV. This is also a vehicle aimed for those who prefer an SUV with a little more style.

Luxury SUVs typically offer the most expected safety features including side airbags, ABS and traction control, and many of them also come with electronic stability control, crash resistant door pillars, dynamic head restraints and back-up sensing systems.

The U.S. News & World Report Rankings and Reviews ranks premium midsize SUVs and crossovers based on an in-depth analysis by its editors of published auto ratings, reviews and test drives. Ranking is based on the score on performance, exterior, interior, safety, and reliability obtained by the vehicles.

Other Names

In Australia and New Zealand, the term *SUV* is not widely used, except by motoring organizations, the press, and industry bodies. Passenger class vehicles designed for off-road use are known as 'four-wheel drives', '4WDs', or '4×4s'. Some manufacturers do refer to their products as SUVs, but others invented names such as XUV, (HSV Avalanche XUV or GMC Envoy XUV) or action utility vehicles (AUVs). The term 'AWD', or all-wheel drive, is used for any vehicle which drives on all four wheels, but may not be designed for off-road use. 'Crossover' is a marketing term for a vehicle that is both four-wheel-drive and primarily a road car.

SUVs such as the Range Rover are sometimes referred to as "Toorak tractors" in Australia, referencing the affluent Melbourne suburb of Toorak

The pejorative term "Toorak Tractor" is used in Australia to describe vehicles such as Range Rovers used in wealthy urban areas with fine roads, fine dining, and exclusive designer shopping precincts where off-road ability is not required. The term alludes to the affluent Melbourne suburb of Toorak and was used at least as early as the late 1980s. The equivalent term "Chelsea Tractor" became prominent in the United Kingdom around 2004 to describe vehicles such luxury SUVs used in urban areas such as Chelsea, London, where their four-wheel-drive capabilities are not required and the car is believed to be a status symbol rather than a necessity. The term "4×4" (four-by-four) is also common even for vehicles not used in urban areas. "AWD" is not commonly used in the UK. The less capable SUVs also pick up the name "soft-roader" because while they appear designed to go off road, in many cases they're not actually capable of it.

In Norway the term "Børstraktor" (Stock Exchange Tractor) serves a similar purpose.

In Finland the term "katumaasturi" is commonly used to designate SUVs. It roughly translates to street-off-roader, or street-4×4. This marks the difference with what is called "maasturi" which is a vehicle with off-road capability.

In Sweden they are often called "stadsjeep" (city jeep).

In the Netherlands they are known as "P.C. Hoofttractors" after the exclusive P.C. Hooftstraat shopping street.

Crossover (Automobile)

Peugeot 2008, a crossover in Europe

A crossover or crossover utility vehicle (CUV) is a vehicle built on a car platform and/or combining, in highly variable degrees, features of a sport utility vehicle (SUV) with features from a passenger vehicle, especially those of a station wagon or hatchback.

Using the unibody construction typical of passenger vehicles instead of the body-on-frame platform used in light trucks and the original SUVs, the crossover combines SUV design features such as tall interior packaging, high H-point seating, high center of gravity, high ground-clearance or all-wheel-drive capability – with design features from an automobile such as a passenger vehicle's platform, independent rear suspension, car-like handling, lighter weight and better fuel economy than trucks or truck-based vehicles.

A crossover may borrow features from a station wagon or hatchback, such as the two-box design of a shared passenger/cargo volume with rear access via a third or fifth door, a liftgate – and flexibility to allow configurations that favor either passenger or cargo volume, e.g., fold-down rear seats.

Crossovers are offered with front wheel drive, rear wheel drive or all-wheel drive configurations. Crossovers are typically designed for only light off-road capability, if any at all.

Sources indicate the term crossover began as a marketing term, and a 2008 CNNMoney article indicated that "many consumers cannot tell the difference between an SUV and a crossover." A January 2008 *Wall Street Journal* blog article called crossovers "wagons that look like sport utility vehicles, but ride like cars."

1974 Neorion Chicago

(1977-current) Lada Niva

1977 Matra-Simca Rancho

1980-87 (MY) AMC Eagle

Though crossovers have become a defined market in recent years, they are not a new concept. The 1948 Willys-Overland Jeepster convertible coupe offered many of the features that define the modern crossover. In 1955, the Russian GAZ-M20 Pobeda was modified into the M-72 version to become the first all-wheel drive car with a unibody. A concept car using a Jeep Wagoneer (SJ) drivetrain was designed in Greece as a luxury limousine and four vehicles were made by Neorion. The Russian off-road Vaz 2121 Niva was introduced in 1976 featuring a unibody body and some mechanical components from the VAZ-2101 compact sedan, the Fiat 124-based Lada, and it saw success as a vehicle "best described as agricultural." Another contender before the crossover description became common was the 1977 Matra Rancho.

A more direct modern crossover antecedent is the AMC Eagle, a passenger road vehicle introduced in 1979 that "pioneered the crossover SUV" category. Featuring "passenger-car comfort, plus 4wd security for all-weather security" the Eagle sedans and station wagons were a contrast to the fuel-thirsty vehicles built for rugged off-road that were marketed in the US at the time. AMC "predicted that consumers would embrace a vehicle with the comfort of an automobile, but the ride height and foul-weather capabilities of a four-wheel-drive utility vehicle." As a precursor to today's crossover models, AMC's "vehicles worked well and sold well" and the "surviving Eagles look like the 'early man' version of a CUV, sort of a missing link of the car world."

The current use of the term for this market segment spans a wide range of vehicles. In some cases, manufacturers have marketed vehicles as *crossovers* simply to avoid calling them station wagons, or have produced crossovers mainly because station wagons have fallen out of favor with buyers in a particular region such as the United States.

While crossover vehicles released in the early-2000s resembled traditional SUVs or wagons, others have prioritized sportiness over utility – such as the Infiniti FX and BMW X6.

By 2006, the segment came into strong visibility in the U.S., when crossover sales "made up more than 50% of the overall SUV market." Sales increased in 2007 by 16%. For Audi, the Audi Q5 has become their second best-selling vehicle in the United States market after the Audi A4 sedan. Around half of Lexus' sales volume come from its SUVs since the late 1990s, the big majority of which is the Lexus RX crossover.

In the U.S., domestic manufacturers were slow to switch from their emphasis on light truck-based SUVs, and foreign automakers developed crossovers targeting the U.S. market, as an alternative to station wagons that are unpopular there. But by the 2010 model year, domestic automakers had quickly caught up. The segment has strong appeal to aging baby boomers.

Examples

The term crossover and SUV are sometimes interchangeable, sometimes used in combination, depending on the marketing or public perception of a particular vehicle. The broad spectrum of crossovers includes, among many others that are marketed in various markets:

- Mini crossovers: e.g., Chevrolet Trax/Opel Mokka/Buick Encore, Citroën C4 Cactus, Mitsubishi Outlander Sport, Fiat Palio Adventure, Ford EcoSport, Honda HR-V, Hyundai Creta, Mini Countryman, Nissan Juke, Peugeot 2008 (#2 in Europe), Renault Captur (#1 in Europe)
- Compact crossovers: e.g., BMW X1, Dacia Duster, Ford Kuga, Jeep Compass, Mahindra XUV500, Nissan X-Trail, Peugeot 3008 (all-wheel-drive Hybrid4 capability), SsangYong Korando
- Mid-sized crossovers: e.g., Acura ZDX, Chevy Equinox/GMC Terrain, Mercedes-Benz M-Class, Ford Edge, Volkswagen Touareg
- Full-sized crossovers: e.g., Buick Enclave, Chevrolet Traverse, Ford Flex, Ford Explorer, Lincoln MKT, Mazda CX-9

The European MPV or large MPV may broadly resemble the crossover, including vehicles such as the Mercedes-Benz R-Class, and Ford S-Max. During the development of the Dodge Journey (Fiat Freemont), Dodge benchmarked several European vehicles.

Sports Car

A sports car, or sportscar, is a small, usually two-seater, two-door automobile designed for spirited performance and nimble handling. The term "sports car" was used in *The Times*, London in 1919. According to USA's Merriam-Webster dictionary, USA's first known use of the term was in 1928. Sports cars started to become popular during the 1920s.

Sports cars may be spartan or luxurious, but high maneuverability and light weight are requisite. Sports cars are usually aerodynamically shaped (since the 1950s), and have a low center of gravity compared to standard models. Steering and suspension are typically designed for precise control at high speeds. Traditionally sports cars were open roadsters, but closed coupés also started to become popular during the 1930s, and the distinction between a sports car and a grand tourer is not absolute

Definition

Attributing the definition of 'sports car' to any particular model can be controversial or the subject of debate among enthusiasts. Authors and experts have often contributed their own ideas to capture a definition.

A car may be a *sporting* automobile without being a sports car. Performance modifications of regular, production cars, such as sport compacts, sports sedans, muscle cars, pony cars and hot hatches, generally are not considered sports cars, yet share traits common to sports cars. Certain models can "appeal to both muscle car and sports car enthusiasts, two camps that rarely acknowledged each other's existences." Some models are called "sports cars" for marketing purposes to take advantage of greater marketplace acceptance and for promotional purposes. High-performance cars of various configurations are grouped as *Sports and Grand tourer cars* or, occasionally, just as *performance cars*.

Drivetrain and Engine Layout

The drivetrain and engine layout significantly influences the handling characteristics of an automobile, and is crucially important in the design of a sports car.

The front-engine, rear-wheel-drive layout (FR) is common to sports cars of any era and has survived longer in sports cars than in mainstream automobiles. Examples include the Caterham 7, Mazda MX-5, and the Chevrolet Corvette. More specifically, many such sports cars have a front mid-engine, rear-wheel drive layout (FMR), with the centre of mass of the engine between the front axle and the firewall.

In search of improved handling and weight distribution, other layouts are sometimes used. The rear mid-engine, rear-wheel-drive layout (RMR) is commonly found only in sports cars—the motor is centre-mounted in the chassis (closer to and behind the driver), and powers only the rear wheels. Some high-performance sports car manufacturers, such as Ferrari and Lamborghini have preferred this layout. The Fiat X1/9 is an example of an affordable mid-engine sports car.

Porsche is one of the few remaining manufacturers using the rear-engine, rear-wheel-drive layout (RR). The motor's distributed weight across the wheels, in a Porsche 911, provides excellent traction, but the significant mass behind the rear wheels makes it more prone to oversteer in some situations. Porsche has continuously refined the design and in recent years added electonic stability control to counteract these inherent design shortcomings.

The front-engine, front-wheel-drive layout (FF) layout which is the most common in sport compacts and hot hatches, and modern production cars in general, is not generally used for sports cars. This layout is advantageous for small, light, lower power sports cars, as it avoids the extra weight, increased transmission power loss, and packaging problems of a long driveshaft and longitudinal engine of FR vehicles. However, its conservative handling effect, particularly understeer, and the fact that many drivers believe rear wheel drive is a more desirable layout for a sports car count against it. The Fiat Barchetta, Saab Sonett, and Berkeley cars are sports cars with this layout.

Before the 1980s few sports cars used four-wheel drive, which had traditionally added a lot of weight. Although not a sports car, the Audi Quattro proved its worth in rallying. With its improvement in traction, particularly in adverse weather conditions, four-wheel drive is no longer uncommon in high-powered sports cars, e.g. Porsche, Lamborghini, and the Bugatti Veyron.

Seating Layout

Traditional sports cars were typically two-seat roadsters. Although the first sports cars were derived from fast tourers, and early sporting regulations often demanded four seats (even three-seaters were often produced by coachbuilders), two seats became common from about the mid-1920s. Modern sports cars may also have small back seats that are often really only suitable for luggage or small children; such a configuration is referred to as a 2+2 (two full seats + two "occasional" seats).

Over the years, some manufacturers of sports cars have sought to increase the practicality of their vehicles by increasing the seating room. One method is to place the driver's seat in the center of the car, which allows two full-sized passenger seats on each side and slightly behind the driver. The arrangement was originally considered for the Lamborghini Miura, but abandoned as impractical because of the difficulty for the driver to enter/exit the vehicle. McLaren used the design in their F1.

Another British manufacturer, TVR, took a different approach in their Cerbera model. The interior was designed in such a way that the dashboard on the passenger side swept toward the front of the car, which allowed the passenger to sit farther forward than the driver. This gave the rear seat passenger extra room and made the arrangement suitable for three adult passengers and one child seated behind the driver. The arrangement has been referred to by the company as a 3+1. Some Matra sports cars even had three seats squeezed next to each other.

Four-wheel Drive

Four-wheel drive, also called 4×4 ("four by four") or 4WD refers to type of a vehicle, specifically one with its drivetrain capable of providing torque to all wheel ends of a two-axled vehicle simultaneously. It may be full-time, or on-demand, and is typically linked via a transfer case which provides an additional output drive-shaft, along with additional gear ranges.

When a four-wheeled vehicle has torque supplied to both axles, this is described as "all-wheel drive" (AWD). However, "four-wheel drive" typically refers to a set of specific components and functions, and/or intended offroad application, which generally complies with modern use of the terminology.

Definitions

4×4/4WD/AWD systems were developed in many different markets and used in many different vehicle platforms. There is no universally accepted set of terminology to describe the various architectures and functions. The terms used by various manufactures often reflect marketing rather than engineering considerations or significant technical differences between systems. SAE International's standard J1952 recommends only the term All-Wheel-Drive with additional sub classifications which cover all types of AWD/4WD/4x4 systems found on production vehicles.

4×4

Four-by-four (4×4) refers to the general class of vehicles. The first figure represents the total wheels (more precisely, axle ends), and the second, the number that are powered. Syntactically, 4×2 means a four-wheel vehicle that transmits engine torque to only two axle-ends: the front two in front-wheel drive or the rear two in rear-wheel drive. Alternatively, a 6×4 vehicle has three axles, two of which provide torque to two wheel ends each.

4WD

Four wheel drive (4WD) refers to vehicles with two axles providing torque to four wheel ends. In the North American market the term generally refers to a system that is optimized for off-road driving conditions. The term "4WD" is typically designated for vehicles equipped with a transfercase which switches between 2WD and 4WD operating modes, either manually or automatically.

AWD

All wheel drive (AWD) historically was synonymous with "four-wheel drive" on four-wheeled vehicles, and six-wheel drive on 6×6s, and so on, being used in that fashion at least as early as the 1920s. Today in North America the term is applied to both heavy vehicles as well as light passenger vehicles. When referring to heavy vehicles the term is increasingly applied to mean "permanent multiple-wheel drive" on 2×2, 4×4, 6×6 or 8×8 drive train systems that include a differential between the front and rear drive shafts. This is often coupled with some sort of anti-slip technology, increasingly hydraulic-based, that allows differentials to spin at different speeds but still be capable of transferring torque from a wheel with poor traction to one with better. Typical AWD systems work well on all surfaces, but are not intended for more extreme off-road use. When used to describe AWD systems in light passenger vehicles, it refers to a system that applies torque to all four wheels (permanently or on demand) and/or is targeted at improving on-road traction and performance (particularly in inclement conditions), rather than for off-road applications.

Some all wheel drive electric vehicles solve this challenge using one motor for each axle, thereby eliminating a mechanical differential between the front and rear axles. An example of this is the dual motor variant of the Tesla Model S, which on a millisecond scale can control the torque distribution electronically between its two motors.

Sedan (Automobile)

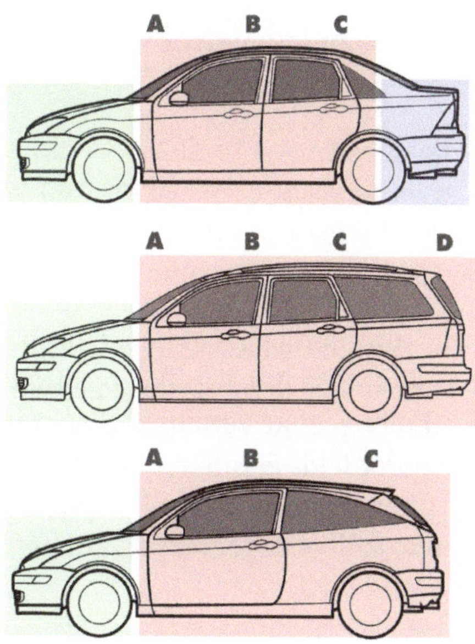

Typical pillar configurations of a sedan (three box), station wagon (two box) |and hatchback (two box) from the same model range

A sedan (American, Canadian, Australian, and New Zealand English) or saloon (British, Irish and Indian English) is a passenger car in a three-box configuration with A, B &

C-pillars and principal volumes articulated in separate compartments for engine, passenger and cargo. The passenger compartment features two rows of seats and adequate passenger space in the rear compartment for adult passengers. The cargo compartment is typically in the rear, with the exception of some rear-engined models, such as the Renault Dauphine, Tatra T613, Volkswagen Type 3 and Chevrolet Corvair. It is one of the most common car body styles. A battery electric sedan such as the Tesla Model S has no engine compartment, but a front cargo compartment and a rear compartment for cargo or optionally for additional passengers.

Types of Sedans

1962 Rambler Classic 2- and 4-door sedans

The primary purpose of the sedan is to transport people and their baggage on ordinary roads. Sedan versions of the automobile body style have a central pillar (B-pillar) that supports the roof and come in two- and four-door versions. Sedans usually have a two-box or three-box body.

In the U.S., the term sedan has been used to denote a car with fixed window frames, as opposed to the hardtop style without a "B" pillar and where the sash or window frame, if any, winds down with the glass. Popular in the U.S. from the 1950s through the 1970s, true hardtop body designs have become increasingly rare.

The shape and position of the automobile greenhouse on both two- and four-door sedans may be identical, with only the center B-pillar positioned further back to accommodate the longer doors on the two-door versions. For example, 1962 Rambler Classic sedans feature identical windshield, A-pillar, roof, C-pillar, and rear window. The two-door sedans have longer doors and include roll down rear side window and even a quarter window that is shaped to follow the reverse slant of the C-pillar, just like on the rear doors of the four-door sedans.

Club Sedans

A two- or four-door design built on a normal chassis, but with a shorter roof and interior space, club sedans were most often available in high-level U.S. models from the mid-1920s to the mid-1950s. Originating from the club car on a Pullman passenger train that was well appointed, the "club" term imparted a sense of class to the smaller-cabin versions in the range of models.

1938 Cadillac Club Sedan

Notchback Sedans

A notchback sedan is a three-box sedan, where the passenger volume is clearly distinct from the trunk volume of the vehicle (when seen from the side). The roof is on one plane, generally parallel to the ground, the rear window at a sharp angle to the roof, and the trunk lid is also parallel to the ground.

1962 Chevrolet Impala, a typical notchback sedan

Fastback Sedans

A fastback sedan is a two-box sedan, with continuous slope from the roof to the base of the decklid (trunk lid), but excludes the hatchback feature. Marketing terminology is often misleading in this area - for example, Daimler AG incorrectly calls the Mercedes-Benz CLS-Class sedan a *four-door coupé* because its semi-fastback design tries to give the impression of a coupé. Certain sedans are edging close to being *one-box* vehicles, where the windshield is steeply raked from the hood/bonnet and the rear window slopes toward almost the end of the car, leaving just a short rear deck that is part of the trunk lid – the 2006 4-door JDM Honda Civic is an example of this. They are not fastbacks because their bodyline changes from the roof to the rear deck. Their steeply raked rear windows end with a decklid that does not continue down to the bumper. Instead, their rear ends are tall – sometimes in a Kammback style – to increase trunk space.

1941 Plymouth fastback sedan

Typically this design is chosen for its aerodynamic advantages. Automakers can no longer afford the penalty in fuel consumption produced by the traditional notchback *three box* form.

Hardtop Sedans

In historic terminology, a sedan will have a frame around the door windows, while the hardtop has frameless door glass. A true hardtop design also has no center or "B" pillar for roof support behind the front doors). This pillarless body style offers greater visibility. However, it requires extra underbody strengthening for structural rigidity. The hardtop design can be considered separately (i.e., a vehicle can be simply called a four-door hardtop), or it can be called a hardtop sedan. During the 1960s and 1970s, hardtop sedans were often sold as *sport sedans* by several American manufacturers and they were among the top selling body styles. During the 1980s, automakers in the U.S. focused on removing weight and increasing strength, and their new four-door sedans with B-pillars were called *pillared hardtops* or *pillared sedans*. The *sport sedan* term has since been appropriated for other uses. In Japan the hardtop design was used for several luxury-type sedans during the 1990s.

1958 AMC Ambassador hardtop

Hatchback Sedans

Chevrolet Malibu Maxx hatchback sedan

Hatchback (a.k.a. liftback) sedans typically have the fastback profile, but instead of a trunk lid, the entire back of the vehicle lifts up (using a liftgate or hatch). A vehicle with four passenger doors and a liftgate at the rear can be called a four-door hatchback, five-door hatchback, four-door hatchback sedan, or five-door sedan. An example of such is the Chevrolet Malibu Maxx and Audi A5 Sportback. There can also be two-door hatchback sedans (three-door sedans), by the same technical explanation for two-door sedans. Examples of this design are the Ford Focus, Ford Fiesta, Volkswagen Golf, Chevrolet Chevette and Daewoo Nexia(Opel Kadett E).

Chauffeured Sedans

The Lincoln Town Car is often used as a chauffeured car in the U.S.

Strictly speaking limousine sedans have a separate compartment for the driver and the passenger compartment is long enough to contain at least two comfortable, forward-facing bench seats. Vehicles used for these means are usually Lincoln Town Car, Cadillac, Mercedes-Benz, or Rolls-Royce.

The term limousine can refer to a large sedan, especially if hired from a service. Chauffeured limousines are primarily used by individuals for weddings, businesses for meetings, as well as for airport and sightseeing transportation. Chauffeurs are professional drivers, usually with experience in the transportation industry or tourism industry.

Chauffeured sedans are owned either by private owners, livery services, or corporations. Large corporations as well as governments commonly provide luxury sedans to top executives, as well as VIP guests. Chauffeured sedans, such as the Lincoln Town Car, may also be stretched into limousines that are capable of seating up to twenty people.

Terminology

Sedan chair carried by two people

The word *sedan* is possibly derived from a southern Italian dialect derivative of Italian *sedia* "chair" (the first sedan was said to have been introduced from Naples). However, Portuguese and Spanish navigators and colonists encountered litters of various sorts in India, Japan, Mexico, and Peru. They were imported into Spain in the late sixteenth century. Soon the fashion spread into France and then England. All the names for these derived from the root "sed-" from the Latin "sella" - the traditional name for a carried chair.

The first automobile to use the configuration was the 1899 Renault Voiturette Type B. The first closed car, for at least 4 persons, which used the word *sedan* was the 1911 Speedwell sedan, which was manufactured by the Speedwell Motor Car Company in Dayton, Ohio. But even before that time completely closed cars were called saloons or limousines, like the 1905 Rational 4-door limousine or the 1907 Renault 4-door limousine or the 1910 Stella 2-door saloon. The words saloon or limousine do not exclusively mean a fully closed car. Cars which are called sedans are almost always fully closed. The term "convertible sedan" was used in the 1930s to describe a car with a soft, foldable top and roll-up windows, very much like a convertible coupe.

The derivation from the town of Sedan, Ardennes in France, where it was said to have been made or first used, lacks historical evidence, according to Oxford English Dictionary.

International Terminology

In American English and American Spanish, the term sedan is used (accented as "sedán" in Spanish). The engine compartment, at the front, is covered by the *hood*; the cargo compartment at the rear is called the *trunk*.

In British English, a car of this configuration is called a saloon. The engine compartment cover is the *bonnet*, the cargo-compartment *boot* is at the rear. Hatchback sedans are known simply as *hatchbacks* (not hatchback saloons); long-wheelbase luxury saloons may be referred to as *limousines*. A sports sedan with hatchback rear are typically called hot hatches in UK, but this usage is largely limited to upgraded small cars.

The British English term *saloon* is sometimes used by British car manufacturers in the United States. For example, the Rolls-Royce Park Ward was sold as a saloon in the United States, while the smaller Silver Seraph was called a sedan.

In Australia the American term sedan is used, albeit with the British terms *boot* and *bonnet* being used. In New Zealand the British terms "boot", "bonnet" and "windscreen" are most commonly used, but the American terms are understood by most of the population. However, the American terms "sedan" and "station wagon" are predominantly used, with "saloon" only finding occasional usage, particularly in the field of motor racing. In other languages, sedans are known as berline (French), berlina (European Spanish, European Portuguese, Romanian, and Italian); although these terms also may include hatchbacks. These terms, besides sedan, derive from types of horse-drawn carriages. In German, the term Limousine is used for sedans, and "Stretch-Limousine" for limousines.

Hatchback

A hatchback is a car body configuration with a rear door that swings upward to provide access to a cargo area. Hatchbacks may feature fold-down second row seating, where the interior can be flexibly reconfigured to prioritize passenger vs. cargo volume. Hatchbacks may feature two- or three-box design.

While early examples of the body configuration can be traced to the 1930s, the Merriam-Webster dictionary dates the term itself to 1970. The hatchback body style has been marketed worldwide on cars ranging in size from superminis to small family cars, as well as executive cars.

Overview

Hatchbacks may be described as *three-door* (two entry doors and the hatch) or *five-door* (four entry doors and the hatch) cars. A model range may include multiple configurations, as with the 2001–2007 Ford Focus which offered sedan (ZX4), wagon (ZXW), and three or five-door hatchback (ZX3 and ZX5) models. The models typically share a platform, drivetrain and bodywork forward of the A-pillar. Hatchbacks may have a removable rigid parcel shelf, liftable with the tailgate, or flexible roll-up tonneau cover to cover the cargo space behind the rear seats.

Volkswagen Polo Mk 1 hatchback

Hatchback Vs. Station Wagon

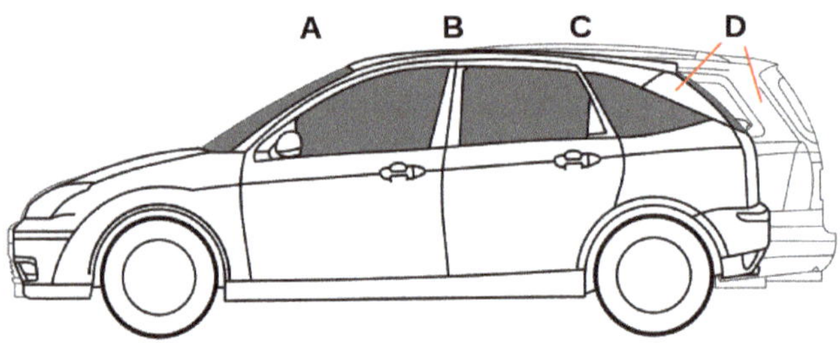

Diagram of a five-door hatchback (two-box) superimposed over the station wagon (two-box) from the same model range—in this case, both with a D-pillar

Both station wagons and hatchbacks typically feature a two-box design configuration, with one shared, flexible, interior volume for passengers and cargo—and a rear door for cargo access. Further distinctions are highly variable:

Pillars: Both configurations typically feature A, B & C pillars; station wagons more likely also feature a D pillar.

Cargo volume: Station wagons prioritize passenger and cargo volume, with windows alongside the cargo volume. Of the two body styles, a station wagon's roof (viewed in profile) more likely extends to the very rearmost of the vehicle, enclosing a full-height cargo volume, whereas a hatchback roof (especially a liftback roof) typically rakes down steeply behind the C-Pillar, prioritizing style over interior volume, and resulting in shorter rear overhang and smaller, or absent, windows either side of the cargo volume.

Cargo floor contour: Favoring cargo capacity, a station wagon may prioritize a fold-flat floor, where a hatchback would more likely allow a cargo floor with pronounced contour (e.g. the new Mini or the sixth generation Ford Fiesta).

Seating: Station wagons have two or three rows of seats (e.g., the Ford Taurus wagons) while hatchbacks have one (e.g. the MGB GT) or two rows of seats.

Rear suspension: A station wagon may include reconfigured rear suspension for additional load capacity and to minimize intrusion into the cargo volume (e.g., worldwide versions of the first generation Ford Focus).

Rear door: Hatchbacks typically feature a top-hinged liftgate for cargo access, with variations from a single liftgate to a complex tailgate that can function either as a full tailgate or as a trunk lid (e.g., the 2008 Škoda Superb's TwinDoor). Station wagons also have numerous tailgate configurations. Typically, a hatchback's hatch or liftgate does not extend down to the bumper, as on wagons. Another appearance variation that seems to blur the lines between a commonly defined hatchback versus a station wagon is called a kammback, which generally features a sloping roof towards the end of the vehicle, with an almost vertical rear section to the bumper.

Automotive journalist Dan Neil, in a 2002 *New York Times* report described verticality of the rear cargo door as the prime distinction between a hatchback and a station wagon: "Where you break the roofline, at what angle, defines the spirit of the vehicle," he said. "You could have a 90-degree break in the back and have a station wagon."

Liftback

A liftback is a broad marketing term for a hatchback where the rear cargo door is more horizontal than vertical, with a sharply raked or fastback profile. In comparison with the hatchback the back opening area is more sloped and longer and is lifted up to open, offering more luggage space. Very similar is the "fastback". Liftback and fastback are often misused. A fastback is described as a hatchback where the rear cargo door flows down continuously from the roof of the car and most of the time is not as raked as the liftback.

Liftback – The 4th generation Toyota Celica, 2.0 SX (ST162)

Early Examples

Citroën Traction Avant c. 1954

1951 Kaiser-Frazer Vagabond

In 1938, Citroën introduced the Citroën Traction Avant in a "Commerciale" version with a tailgate, initially with a two-piece tailgate, of which the upper piece hinged upwards, cutting well into the roof, and after 1954 with a one-piece top-hinged hatchback.

In 1946, DeSoto marketed the Suburban as a station wagon, but it was an extended sedan with a trunk lid that was hinged below the rear window. The model was promoted as offering station wagon utility with the passenger and luggage compartments in one large section with an 8-foot (2,438 mm) reasonably flat floor and equipped with folding rear seats to provide flexibility similar to hatchbacks.

In 1949, Kaiser-Frazer introduced the Vagabond and Traveler hatchbacks. Although these were styled much like the typical 1940s sedan, they incorporated an innovative split rear tailgate, folding rear seats, and no separate trunk. The design was neither fully a sedan nor a station wagon, but the folding rear seat provided for a large, 8-foot (2.4 m) long interior cargo area. These Kaiser-Frazer models have been described as "America's First Hatchback".

In 1953, Aston Martin marketed the DB2 with a top-hinged rear tailgate, manufacturing 700 examples. Its successor, the 1958 DB Mark III, also offered a folding rear seat. The 1954 AC Aceca and later Aceca-Bristol from AC Cars had a similar hatch tailgate, though only 320 were built.

The British Motor Corporation launched a 'Countryman' version of the Austin A40 Farina compact car in 1959, which incorporated a horizontal-split two-piece tailgate and a more vertical rear panel in comparison to the standard saloon version. This was close to the enduring concept of the hatchback: A small car with a large rear door aperture providing a versatile combination of rear passenger space and easy loading for cargo, although with a two-piece rear door it was not quite a true hatch.

Worldwide

1961 Renault 4, the world's first million selling hatchback

Europe

In 1961, Renault introduced the Renault 4, a small car with a top-hinged tailgate incorporating the rear window, and only short side windows between C & D-pillars aside the luggage space and a steep angle from roof to rear bumper. Although during its production run the R4 was called a small station wagon -even after the term *hatchback* appeared around 1970 - with over eight million cars built, this was the first mass-produced hatchback. Despite the success of subsequent smaller hatchback models including the Renault 5, the R4 continued in production in Europe until 1986 and through 1992 in Argentina.

1964: With the Renault 16, hatchbacks became an example for compact cars

Peugeot 306 hatchback, with the hatch lifted

In 1964, Fiat subsidiary Autobianchi marketed the Primula hatchback, replaced by Autobianchi A112 in 1969. In Poland, a FSO Syrena prototype "110" small hatchback was developed in 1964.

In 1965, Renault marketed the Renault 16, a mid-market hatchback design with a folding rear seat. It was designed to compete across Europe with traditional family saloons including Britain's Ford Cortina, Vauxhall Victor and Morris Oxford. However, Renault continued to produce saloon alternatives to the R16, which remained in production until 1979.

Also in 1965, the MGB-GT was launched with a hatchback designed by Pininfarina, the first volume-production sports car so equipped. In 1966 Peugeot offered a shortened three-door coupe version of their 204, essentially a hatchback. The same year saw Innocenti, which built the Austin A40 under license in Italy, introduced the A40 Combinata with a top-hinged one-piece tailgate, making the car into a true hatchback.

In 1967, the Simca 1100 used a transverse engine and gearbox layout, and incorporated a hatchback without side windows at the C-pillar. In the same year, there appeared more conventional Citroën Dyane.

In 1969, British Leyland launched Austin Maxi, a five-speed, transverse front-wheel drive hatchback. This was similar in size to the Renault 16, although saloons remained the most popular choice of bodystyle in this sector across Europe for more than a decade afterwards.

The Fiat 127 was launched as a hatchback in 1971, as was the Renault 5 a year later, with these models popularising the hatchback bodystyle on Europe's smaller cars. In Yugoslavia, in 1971 there started production of Zastava 101.

In 1973, Volkswagen marketed the Passat/Dasher hatchback (similar in size and concept to the Renault 16 and Austin Maxi), followed by the Golf/Rabbit designed by Giorgetto Giugiaro, as well as the Audi 50/Volkswagen Polo in 1974. The Golf became a classic compact hatchback and its newer generations have remained in production until now.

General Motors followed in 1975, when it made use of the hatchback bodystyle for the first time on its Chevette (Opel Kadett City in Continental Europe).

Backed by the great success of the Renault 16, in 1975, Renault decides to launch the 20 and 30 as its two first executive cars produced until 1984, making them the first French luxury cars with hatchbacks.

Sports cars like the Jaguar E-Type with its side-hinged opening, Toyota 2000GT, and Datsun 240Z carried rear tailgates, with one row of seats, and Ford relaunched the Capri with a hatchback in 1974. In the 1970s, the Rover SD1, Renault 30, and Saab 900 introduced the hatchback style into the executive car market.

By the late 1970s and early 1980s, the majority of superminins and compact cars had been updated or replaced with hatchback variations, such as the 1978 Fiat Ritmo/Strada and FSO Polonez, the 1979 Citroën GSA, Lancia Delta and Opel Kadett (Vauxhall Astra in Britain) and the 1980 Ford Escort and Austin Metro.

The 1983 Fiat Uno designed by Giorgetto Giugiaro, its tall, square body utilising a Kamm tail achieved a low drag coefficient of 0.34 won much praise for airy interior space and fuel economy. It incorporated many packaging lessons learnt from Giugiaro's 1978 Lancia Megagamma concept car (the first modern people carrier / MPV / mini-van) but miniaturised. Its tall car / high seating packaging is imitated by every small car today. It reversed the trend for lower and lower built cars. It showed that not just low sleek cars could be aerodynamic, but small, roomy, boxy well packaged cars could be too. In 1984 it was voted European Car of the Year.

In 1983, the Renault 25 replaces the 30 as its executive model utilizing the hatchback which was very popular at the time in southern Europe.

General Motors launched its new "J Car" range in 1981, produced under various marques across the world, to compete in the larger family car sector. The Continental version was sold as the Opel Ascona, while the British version was the Vauxhall Cavalier. A saloon version was produced alongside the hatchback, a bodystyle not seen previously on any Opel or Vauxhall of this size. A year later, Ford called time on production of the conventional Cortina saloon after 20 years and five incarnations to switch to a hatchback bodystyle for the Sierra.

In early 1983, Austin Rover moved into the medium-sized hatchback market popularised by the Volkswagen Golf with its new Austin/MG Maestro. A saloon version, the Montego, was launched a year later to compete with the likes of the Ford Sierra and Vauxhall Cavalier.

Following the executive trend, in 1984, Saab proposed the 9000 as part of the Type Four platform in conjunction with the Italian automaker Fiat Automobiles. Along with the FIAT Croma, Saab's model was featuring the hatchback.

Traditional saloons and estates also remained popular throughout the 1980s, and some manufacturers sold their hatchback-based saloons under different names. Renault had launched the R9 saloon in 1981; this spawned a hatchback, the R11, in 1983. Ford had marketed the MK3 Escort as a hatchback and estate from 1980, but the saloon version launched in 1983 was badged as the Orion. Since 1980, Volkswagen had produced a Golf-based saloon called the Jetta. The MK2 Vauxhall Astra launched in 1984 was soon followed by a saloon version called the Belmont. When Ford updated its Sierra in early 1987, a saloon version was made available for the first time and marketed as the Sierra Sapphire.

The hatchback offered practicality for consumers in Europe. From the 1960s it was adopted as a standard feature on most European cars, with saloons declining in popularity apart from at the top of the market where a saloon is seen as a sign of status. The Ford Granada Mk3 /Ford Scorpio executive car, that was launched in 1985 only as a hatchback, was joined by a saloon in 1990 and an estate in 1992. When the final Scorpio was launched in 1995, it was sold as only a saloon or estate. General Motors never produced a hatchback version of any of its top of the range Vauxhall or Opel models, while Rover had discontinued the hatchback bodystyle in this sector by 1999, while Renault continued the bodystyle as late as 2009 with the VelSatis.

BMW added a hatchback to its range in 1968 with the BMW 1802-2002 Touring and again in 1994 with the E36 3-series hatchback, and Mercedes-Benz in 1997 with the A-Class and the Sportcoupé in 2000. Audi returned to the hatchback market with the Audi A2 in 1999.

In the 2010s, executive hatchbacks found a resurgence after Renault Vel Satis was dropped in 2009, with the BMW 5 GT and Porsche Panamera launched later the same year, followed by the Audi A7 in 2010.

North America

American Motors Corporation (AMC) marketed the subcompact Gremlin from 1970, in a single hatchback Kammback body design. The Gremlin used the AMC Hornet automobile platform, but its abrupt hatchback rear end cut the car's overall length from 179 to 161 inches (4,500 to 4,100 mm). American Motors added a semi-fastback hatchback version to its larger compact-sized Hornet line for the 1973 model year. The design and fold-down rear seat more than doubled cargo space and the Hornet was claimed to be the "first compact hatchback" manufactured by U.S. automaker. Additional utility from the car's hatchback, space, and long load floor was that it could be turned into a "mini-camper" with the optional factory "Hornet Hutch" tent accessory.

AMC Hornet hatchback with mini-camper accessory

Introduced by AMC in 1975, "like recent European variations of the theme, the Pacer had a rear door or hatchback, which further increased its utility". For 1977, AMC added a longer Pacer model with a wagon-type configuration describing its large rear "hatch" as one of the car's three doors, all having different sizes. The Hornet's hatchback body design was continued in the redesigned "luxury" Concord line for 1978 and 1979, in a "sporty model designed for performance-oriented buyers". The AMC Spirit replaced the Gremlin starting with the 1979 models and was available in two designs, both featuring rear doors: a hatchback "sedan" and a semi-fastback "liftback" version.

Built in AMC's Kenosha, Wisconsin factories, the 1984–1987 Encore was a two and four-door hatchback based on the European Renault 11.

Chrysler Corporation introduced the Dodge Omni and Plymouth Horizon hatchbacks in 1978 which were modeled that were designed as the Simca Horizon in France. These were followed by the Dodge Charger/Plymouth Turismo. They released the liftback Dodge Daytona/Chrysler Laser in 1984, and then the Dodge Shadow/Plymouth Sundance—as well as the LeBaron GTS and Dodge Lancer hatchbacks. Captive import subcompact models included the Dodge Colt and the almost identical Plymouth Champ.

Ford Motor Company introduced the Pinto and Bobcat in 1972. The German-built Mercury Capri II hatchbacks were imported to U.S. Lincoln-Mercury dealers for the 1976–1977 model years, and the Ford Fiesta hatchback was imported for U.S. Ford dealers later in the decade. Ford offered a hatchback version of its third-generation Escort. The third generation Mustang and Mercury Capri introduced in 1979 were offered in hatchback models. Between 1988 and 1993 Ford marketed the imported Festiva subcompact hatchback that was later restyled and renamed the Aspire for the 1994 through 1997 model years.

1973 Chevrolet Vega Hatchback

The Chevrolet Vega, introduced in September 1970, was the first hatchback model from General Motors. Over a million Vega hatchbacks were produced for the 1971–1977 model years accounting for about half of the Vega's total production. GM introduced rebadged Vega hatchback variants, the 1973–1977 Pontiac Astre and the 1978 Chevrolet Monza S.

The Vega-derived Chevrolet Monza 2+2, Buick Skyhawk, and Oldsmobile Starfire introduced for the 1975 model year, were produced exclusively as hatchbacks with the Pontiac Sunbird hatchback introduced for the 1977 model year. All were produced through 1980.

A Chevrolet Nova hatchback was introduced for the 1973 model year, and was offered through 1979, another hatchback Nova was reintroduced in the 80's based on the Toyota Sprinter. The Chevrolet Chevette was introduced in 1975 as a two-door hatchback, engineered in cooperation with Isuzu. A four-door hatchback on a longer wheelbase was introduced with the 1978 models. In early 1979 the Chevrolet Citation was introduced as a 1980 model offered in 2 and 4-door hatchbacks continuing through the 1987 model year. In the 1981, General Motors included a hatchback model as part of its "J-body" platform that included the Chevrolet Cavalier. Chevrolet offered captive import hatchbacks built by Suzuki and Izuzu. The NUMMI U.S.-made Chevrolet Nova was also offered in a hatchback model in 1987 and 1988. Its replacement, the Geo Prizm, was also available in a hatchback model and the domestic designed Chevrolet Corsica was briefly available in a hatchback version.

The third generation Chevrolet Camaro and Pontiac Firebird produced for the 1982–1992 model years, featured a curved glass hatchback liftgate. GM marketed a series of hatchbacks in North America as a joint venture with Suzuki, the Swift/Metro/Firefly. Chevrolet offered a longer wheelbase, hatchback version of the Malibu, the Malibu Maxx from 2004 to 2007. In 2008, GM introduced the 3-door and 5-door Belgian-assembled Saturn Astra. Chevrolet added a hatchback version of its Korean-built Aveo in 2009.

The Chevrolet Corvette was first offered with an opening rear glass hatch for the 1982 Collector Edition model. It was adopted on all Corvette Coupes beginning in 1984, with the fourth generation models.

Japan

Nissan Fairlady Z (S30)

One of the first hatchbacks offered was the 1970 Nissan Fairlady Z, 1972 Honda Civic, the Nissan Sunny, and the Nissan Cherry, with the Civic and Cherry offering front wheel drive powertrains. In addition to specific models of captive imports mentioned above, a number of Japanese brands have been available in the hatchback body style as a primary model: Nissan Pulsar, Toyota Allex, Toyota Prius, Honda CR-X, and the Honda Insight. Almost all Japanese "city cars", called "kei jidosha" use a hatchback bodystyle for cargo carrying ability in a regulated vehicle size, such as the Mitsubishi Minica, Honda Life, Suzuki Fronte, Subaru Vivio, and Daihatsu Mira. Other large-sized hatchback body style vehicles include Lexus CT, Nissan Murano, and the Nissan Skyline Crossover. The Nissan Fairlady Z is often classified as hatchbacks because of its styling, including the Z33 and the Z34. Other sports cars with hatchback (liftback) body style in Japan are Toyota Celica and Nissan 180SX.

USSR

Lada Samara 2109

The first Soviet hatchback was the rear-wheel drive IZh 2125 Kombi, which entered production in 1973. This was followed only in the 1980s by the front-wheel drive Lada Samara in 1984, the Moskvitch 2141/Aleko in 1986 and ZAZ Tavria in 1987.

India

Maruti 800 hatchback in India

Hatchbacks are popular in India. The Maruti 800 sold over 2.5 million units since its launch in 1983. Since 2004, Maruti 800 has been overtaken by Maruti Alto as the car with highest annual sales. In March 2009, Tata Motors launched the Nano hatchback, the least expensive road car in the world.

Other Regions

Hatchbacks have proved to be less popular in South America, Africa, and some parts of Asia than in Europe, and as a result, manufacturers have had to develop sedan versions of their small cars. In Brazil, for example, the Fiat Premio was developed from the Fiat Uno in the 1980s, with Ford and GM subsequently offering sedan versions of the Opel Corsa and Ford Fiesta in the 1990s. (The first generation Opel Corsa was sold in Europe as a sedan as well as a hatchback, but proved unpopular, and the three-box sedan was not replaced in 1993). These models were also sold in South Africa and China.

Luxury Vehicle

A Rolls-Royce model Phantom IV, a 1950s luxury car

Luxury vehicle is a marketing term for a vehicle that provides luxury — pleasant or desirable features beyond strict necessity — at increased expense.

The term suggests a vehicle with higher quality equipment, better performance, more precise construction, comfort, higher design, technologically innovative modern, or features that convey an image, brand, status, or prestige, or any other 'discretionary' feature or combination of them. The term is also broad, highly variable and relative. It is a perceptual, conditional and subjective attribute that may be comprehended differently by different people; "What is a luxury car to some... may be 'ordinary' to others."

In contemporary usage, the term may be applied to any vehicle type— including sedan, coupe, hatchback, station wagon, and convertible body styles, as well as to minivans, crossovers, or sport utility vehicles and to any size vehicle, from small to large—in any price range. Moreover, there is a convergence in the markets and a resulting confusion of luxury with high price: where there may have been a clear difference in price between luxury and others, there is no longer an absolute separation between premium and luxury, with what may be premium brands now more expensive than the equivalent so-called luxury ones.

Definition

A Duesenberg, "one of the great luxury cars" with custom body by Willoughby

Automobile manufacturers market specific makes and models that are targeted at particular socio-economic classes, and thus "social status came to be associated more with a particular vehicle than ownership of a car per se." Therefore, automakers differentiate among their product lines in "collusion" with the car-buying public. While a high price is the most frequent factor, it is "styling, engineering, and even public opinion which cars had the highest and lowest status associated with them."

Every era in automobile history has had "a group of car marques and models that have been expensive to purchase, due to their alleged superiority of their design and engineering". Aimed at wealthy buyers, such automobiles might be generically termed *luxury cars*. This term is also used for unique vehicles produced during "an era when luxury was individualistic consideration, and coachwork could be tailored to an owner like a bespoke suit." Although there is considerable literature about specific marques, there is a lack of systematic and scholarly work that "analyzes the luxury car phenomenon itself."

Luxury vehicle makers may either be stand-alone companies in their own right, such as BMW and Mercedes-Benz, or a division/subsidiary of a mass market automaker (e.g., Lexus is part of Toyota). Badge engineering is often used for cost savings, for example, the Lincoln vehicles that are based on Ford platforms or Acura models derived from Honda.

Though widely used, the term luxury is broad and highly variable. It is a perceptual, conditional and subjective attribute and may be understood differently by different people: "What is a luxury car to some... may be 'ordinary' to others."

Global References

Italian Isotta Fraschini Tipo 8A S LeBaron Boattail Roadster, a 1930s luxury car

According to the European Commission, the "luxury vehicle" segment is classified as F-segment. However, the boundaries between the traditional segments are increasingly becoming blurred and diluted as features once exclusive to luxury vehicles become standard equipment on even small cars.

- ACRISS is a code system used by many car rental companies in the US for classifying vehicles (but not brands or marques). The system includes "Luxury" and "Premium" categories.
- Australia: In Australia, for taxation purposes a luxury car is defined as a passenger car whose value exceeds a certain threshold.
- France: In France the term "voiture de luxe" is used.
- Germany: In Germany the term "Upper class" (German: *Oberklasse*) is used.
- Russia: Russian markets use the term "representative class vehicle" (*автомобиль представительского класса*, also translated as "luxury vehicle").

- United States: The *Consumer Guide* publication uses a categorization that sorts luxury vehicles by size and it acknowledges that there can be considerable price variations within a class; for instance the Lincoln Town Car has a relatively low MSRP (Manufacturer's Suggested Retail Price) compared to the Jaguar XJ, BMW 7 Series, Lexus LS, or Mercedes-Benz S-Class.

Luxury Market Segments

Premium Compact Segment

The premium compact class is relatively new, having been initiated by several European brands in the mid-2000s, and have displaced their compact executive cars to constitute the least expensive offerings in their lineups. The classification varies, for instance Consumer Guide Automotive in the U.S. considers the Audi A3 and Mercedes-Benz CLA-Class as part of the premium compact segment due to similar size and MSRP, though these are known in Europe as a small family car and a compact executive car, respectively. Adding to some confusion, General Motors has positioned Buick as a "premium" nameplate to compete with Lincoln (whereas it formerly competed with Mercury) and front-wheel drive Acura and Lexus models, while Cadillac is aimed at the "luxury" segment which generally encompasses the performance-oriented BMW and Mercedes-Benz marques.

The Audi A3 is an example of a premium compact car.

The premium compact segment is targeted at a niche market of consumers who found the existing entry-level luxury offerings (mostly consisting of compact executive cars to be too expensive. By offering a smaller, lighter, more fuel-efficient, and less expensive vehicle, premium compacts introduces younger buyers to the luxury marque, in hopes of retaining the coveted customer loyalty. This includes the Acura CSX, Audi A3, Buick Verano, BMW 2 Series, BMW i3, Cadillac ELR, Chrysler 200, Lexus CT, Mercedes-Benz CLA-Class, Mercedes-Benz B-Class, Saab 9-2X, and the Volvo V40. Premium compacts compete with well-equipped midsize (non-luxury) cars, and with options they overlap much with compact executive cars (entry-level luxury cars).

Premium compacts may share components with mass market cars from the marque's parent company (the Audi A3 for example, shares much of its engineering with the cheaper Volkswagen Golf), and/or have less sophisticated platforms compared to up-market vehicles in the lineup (such as the Mercedes-Benz B-Class which is front-wheel drive compared to the other marque's more expensive cars that are rear-wheel drive). However, the BMW 1 Series retains the front-engine longitudinal-engine rear-wheel drive configuration of more expensive BMWs.

The body style tends to be a hatchback or compact wagon, previously associated with economy cars but regaining popularity in the United States for its afforded utility. As of 2013 there are small luxury sedans such as Audi A3, and Mercedes-Benz CLA-Class. The luxury branding and style, high-quality interior materials, wide range of convenience features, and performance powertrains are key to distinguishing them from mass market equivalents (one mistake made by the Cadillac Cimarron) and making these appeal to consumers.

Entry-level Luxury/Compact Executive Cars

The Lexus IS is an example of an entry-level luxury car

This category is known as the compact executive car in Britain where it specifies both the price range and vehicle dimensions. In the U.S., there is a broader category called entry-level luxury, which includes the bottom vehicles in the line-up of luxury brands as well as the top-of-the-line models of some non-luxury brands. Dimensionally, compact executive cars are smaller than Mid-size cars and sometimes even smaller than compact/Small family cars. Mass market compact cars typically use the economical front wheel drive transverse engine layout, well suited to the inline-4 engine, which also maximizes interior room. Most compact executive cars are rear-wheel drive with longitudinal engines, for improved stability and handling, and in order to accommodate the larger size of higher-performance engines (straight-6, V6, rarely V8), with four-wheel drive often being available. Compact executive cars also tend to have more complicated independent suspensions, sportier transmissions, and high revolution engines that may require premium gasoline. The more complex powertrain and mechanical layouts of compact executive cars comes at increased cost and reduced interior passenger and trunk space. Compact executive cars include the Acura ILX, Alfa Romeo Giulia, Jaguar XE, Volvo S60, Lexus IS, Infiniti Q40, Saab 9-3, BMW 3 Series, Audi A4, Cadillac ATS, and Mercedes-Benz C-Class which particularly emphasize sporty handling.

In the United States and Canada these vehicles, and larger premium offerings at the same market position, are known as entry-level luxury cars.

2nd generation Buick LaCrosse, example of a large American entry-level luxury car.

The American and Canadian classification of entry-level luxury (formerly dominated by premium marques e.g. Buick, Oldsmobile, Mercury, Chrysler) is a broad category which includes not only compact executive cars, but also midsize to large cars and SUVs such as the Lexus ES, Acura TL, Buick LaCrosse, Lexus RX being similar in price to the BMW 3 Series and Mercedes-Benz C-Class. However, the ES, TL, and LaCrosse do not fall under the European luxury classification of compact executive car nor are they sold in Europe. The ES and TL actually began as badge engineered versions of the mid-size Toyota Camry and Honda Accord, respectively, retaining the front wheel drive transverse engine layout, whereas Toyota's current Lexus IS does not share a platform with mass market Toyota cars. The ES and TL are not considered executive cars either (that slot is filled by the Lexus GS and Acura RL, respectively), despite similar dimensions, since executive cars usually emphasize higher performance, often being rear-wheel drive with longitudinal engines. The present iterations of the ES and LaCrosse emphasize "comfortable, reasonably priced luxury", while the TL has recently been targeted at the entry-level luxury sport market due to its SH-AWD.

Recently, the entry-level luxury has been very competitive, and there has been price-overlapping with well-equipped non-luxury cars. The bestselling vehicles of the marque are frequently compact executive cars, such as the BMW 3 Series that makes up 40% of the vehicles that BMW sells worldwide.

Mid-size Luxury/Executive Cars

The mid-size luxury segment is commonly referred to as executive cars in Britain, *Obere Mittelklasse* in German, and *Grandes Routières* in French. Examples of models in this category include Acura RLX, Audi A6, Audi A7, Volvo S80, Volvo S90, Genesis G80, BMW 5 Series, Mercedes-Benz E-Class, Jaguar XF, Maserati Ghibli, Cadillac CTS, Lincoln MKZ, Saab 9-5, Lexus GS, and Infiniti Q50. Although Maserati vehicles have traditionally been priced as ultra-luxury cars, the new Maserati Ghibli III is intended to compete with high-spec executive cars, such as the Audi S6/S7, BMW 550i and Mercedes-Benz E550/CLS550.

The Hyundai Genesis is example of a large/mid-size luxury sedan sold initially under a non-luxury brand, but later on becoming the corner stone of a new luxury marque.

The BMW 5-Series is an example of a mid-size luxury sedan.

Although having similar dimensions to mid-size cars and large family cars, executive cars are engineered and positioned as premium vehicles with better performance and technology amenities. There are also higher development and production costs, as many well-known mid-luxury cars use the longitudinal (usually front engine rear wheel drive layout) engine mounting which affords space for powerful engines such as a V8, instead of the more economical transverse engine layout of mass market cars which are restricted to inline-4 or V6. Executive cars usually have lower sales and production volumes compared to mid-size cars, as base trims with less equipment and smaller engines are not sold in U.S. and Canadian markets. Also due to the problem of steep depreciation, especially concerning cars from less prestigious brands, most executive cars are produced by marques that specialize in larger/more expensive vehicles. Particularly in the United States, the mid-luxury segment is dominated by Mercedes-Benz and BMW, as executive cars from lesser-known manufacturers such as Citroën and Infiniti have not had much success.

Vehicles in this segment include the mid-range models of several luxury car manufacturers. Executive cars such as the BMW 5 Series are crucial to a luxury automaker's bottom line, and although not the highest-selling model, they generate a significant amount of profits due to the lucrative technology options.

High-end Luxury/Full-size Luxury Cars

The Maserati Quattroporte is an example of a full-size luxury car

Also known as full-size luxury cars, grand saloons, flagship car, or premium large cars, while "Oberklasse" is used in Germany. Many of these are the marque's showcases for the newest automotive technology. Several nameplates also offer long-wheelbase versions that offer additional rear legroom and amenities. Full-size luxury cars sold in Canada and the U.S. typically have mid-displacement V8 engines, though recently some marques have offered six cylinder powerplants but without much success.

Vehicles in this category include some of the models from the flagship lines of luxury car brands. In China: Hongqi L5, Emgrand GE, & Hongqi HQE. In Europe, this includes the Audi A8, BMW 7 Series, Porsche Panamera, Volkswagen Phaeton, Maserati Quattroporte, Mercedes-Benz S-Class, and the Jaguar XJ. In Japan: Acura RLX, Lexus LS, Nissan Cima, Nissan President, Toyota Century, Toyota Crown, Toyota Crown Majesta, and the Infiniti Q70. In Australia: HSV Grange / Holden Caprice. In South Korea:SsangYong Chairman, Hyundai Equus, Genesis G90, Kia K900, in the US: Lincoln Continental, Chrysler 300, Tesla Model S, Cadillac XTS, and the Cadillac CT6. In the United States in 2004, the bestselling full-size luxury nameplates were the Lexus LS, Mercedes-Benz S-Class, and BMW 7 Series. There were also full-size luxury cars in the Soviet Union since 1930s until the 1990s, from the GAZ and ZiL (before 1956 - ZiS) factories, such as ZIS-101, ZIS-110, GAZ 12, GAZ 13 Gull, ZIL-114, ZIL-117, ZIL-4104, ZIL-41047.

Ultra-luxury Cars

The Bentley Mulsanne is an example of an ultra-luxury car

R.L. Polk and Company, a global automotive information and marketing firm that provides services to automotive and related industries, has defined the term "super luxury" segment for luxury cars costing more than US$100,000.

This bracket includes the full lineups of Aston Martin, Rolls-Royce and Bentley, as - unlike mainstream luxury brands such as Audi, Porsche, BMW, and Mercedes-Benz - these marques do not offer lower-priced, entry-level luxury or mid-luxury cars. As of late 2011, super luxury automakers have been "pledging higher sales and offering ever sexier models" despite the economic uncertainty.

Bentley and Rolls-Royce have recently moved into more affordable price brackets with new models priced considerably lower than their traditional offerings (Bentley Mulsanne and Rolls-Royce Phantom and Maserati Quattroporte). These new cars, the Bentley Flying Spur and Continental GT sister cars and Rolls-Royce Ghost, created what *Car and Driver* described as the "entry-opulent segment". Bentley, Maybach, and Rolls-Royce vehicles share platforms and derivatives of engines with other luxury brands from their parent auto company, however Rolls-Royce and Bentley are assembled in England (separate from the rest of BMW and Volkswagen Group's production plants) and this 'exclusivity' has helped to make these British marques a sales success. By comparison, Maybachs were built alongside the Mercedes-Benz S-Class which partly explains why they have not fared well in the market and were discontinued in 2012. Furthermore, the Maybach's brand pedigree was virtually unknown outside of Germany unlike its British rivals which have long enjoyed fame worldwide. A Bentley executive was quoted as saying "that the brand's exclusivity, history and obsessive luxury help to convince customers that a Bentley is worth the price" which is at least twice that of a flagship luxury car from BMW or Mercedes-Benz. Although Rolls-Royce and Bentley have traditionally been considered "as a purveyor of boxy British land yachts", newer offerings such as the Bentley Continental Flying Spur and Bentley Continental GT have "upended the super-premium auto segment, making Bentley for the first time a plausible choice for younger Hollywood and music stars".

Maserati vehicles such as the Maserati Quattroporte have traditionally been priced as ultra-luxury cars, although the new Maserati Ghibli III is intended to compete with high-spec executive cars, such as the Audi S6/S7, BMW 550i and Mercedes-Benz E550/CLS550.

The most expensive variants of flagship cars from Audi, BMW, and Mercedes-Benz are often included here as well. They often feature a longer wheelbase and have V12 engines for great refinement and smoothness, even though contemporary V8 engines can make similar output while being less expensive. While BMW sells far fewer V12-engined 7-Series vehicles than V8 versions, the V12 retains popularity in the US, Brazil, China, and Russia, as well as maintaining the marque's prestige in the luxury vehicle market segment. These trims also feature standard luxury and technology features that would be considered optional on their V8 siblings.

There are ultra-high performance cars from "exotic brands" that also exceed US$100,000, but would not necessarily be categorized as luxury automobiles, such as the sports cars from Ferrari, Lamborghini, and Porsche, although the Porsche Panamera does compete in the high-end luxury/full-size luxury category.

Luxury SUV/Crossover

Although only becoming popular in the 1990s, this category has been said to be created with Kaiser Jeep's 1966 Super Wagoneer as the first true luxury four-wheel drive (4x4) vehicle. It was the first serious off-road SUV to offer a V8 engine, automatic transmission, and luxury car trim and equipment. The Super Wagoneer was "a pioneer that blazed a trail for today's luxury SUVs". It came with long list of standard equipment that included bucket seating with center console, air conditioning, seven-position tilt steering wheel, a vinyl roof, as well as "Antique Gold" trim panels on the body sides and tailgate. After American Motors Corporation (AMC) purchased Jeep, the vehicles were upgraded and refined, including features such as an optional electric sliding steel sun roof, "possibly the first offered on an SUV". The late-1970s Jeep Wagoneer Limited "set the sport-utility market on its ear ... was the most luxurious four-wheeler anyone had ever seen."

The Cadillac Escalade is an example of a full size luxury SUV

The Mercedes-Benz GL-Class is an example of a luxury SUV/Crossover

The venerable Grand Wagoneer was still the leader in the luxury SUV market in 1986, and continued to be produced virtually unchanged through 1991, long after Chrysler bought out American Motors Corporation (AMC) in 1987. The SUV market soon expanded with new entrants. By the mid-1990s, the U.S. Big Three automakers, "especially with Ford's Explorer and Jeep's Grand Cherokee, dominate[d] the market for the red-hot sport/utilities." The fastest growing sector of this market was for the so-called luxury SUVs, which included the Jeep Grand Cherokee ... the Grand Cherokee's allure: "This vehicle is proof you can have a true off-road vehicle without giving up luxuries and amenities" with the Jeep providing a crucial new intangible factor for buyers — image.

The SUV models generated higher-profit-margins than ordinary automobiles, and automakers introduced new luxury models during the late 1990s, starting with Lincoln Navigator in 1997, besides traditional models like Range Rover. For some manufacturers such as Porsche and BMW, luxury SUVs were the first SUV models they produced. Luxury SUVs catered particularly to the U.S. market where station wagons were unpopular, often being produced in North America (such as BMW Spartanburg) instead of the luxury marque's home country. Some of these models were not traditional SUVs based on light trucks, rather they are classified as crossovers using unibody constructions. SUVs from non-luxury brands had experienced a surge in popularity through early 2000s, causing the traditional luxury marques to follow.

SUVs from the luxury marques grew at almost 40 percent to more than 430,000 vehicles, excluding SUV-only brands like Hummer and Land Rover, while luxury car sales in the U.S. during 2003 suffered a 1% decline, and non-luxury SUV sales were flat. By 2004, 30 percent of major luxury brands' U.S. sales are now SUVs. Luxury brands in particular led the development of crossover SUVs (as opposed to body-on-frame SUVs), making it one of the fastest growing segments in the market, as the forecast for 2002 was approximately 240,000 vehicles and that could double by 2006. Research data showed luxury SUV buyers are compared those vehicles to SUVs of mass market brands, and not shopping around luxury cars, thus the SUV is becoming the key to bringing new customers to the luxury dealerships.

Certain luxury SUVs use body-on-frame underpinnings, often being badge-engineered versions of their non-luxury counterparts, and retaining the rugged off-road and towing capabilities. Examples include the Lexus LX and Lincoln Navigator, which are the premium versions of the Toyota Land Cruiser (which is considered as a luxury SUV itself) and Ford Expedition, respectively.

Other luxury SUVs are crossovers using unibody construction, often sharing the platform with compact executive and executive cars. For example, the Infiniti FX is based upon the Nissan FM platform that also underpins other Infiniti cars. Audi and BMW developed crossovers to compete in the SUV segment as they did not have an existing body-on-frame vehicle in their lineup. The Lexus RX was the earliest luxury crossover on the market, and it has since been the best-selling luxury vehicle in the US, so it has inspired similar competitors from rival marques. While early luxury crossovers released in the late 1990s have resembled traditional boxy SUVs, recent offerings have prioritized sportiness over utility – such as the Infiniti FX and BMW X6.

Characteristics

A Spanish luxury car brand was Hispano-Suiza, here is twelve cylinder J12

Features

Luxury cars tend to offer a higher degree of comfort than their mainstream counterparts, common amenities include genuine leather upholstery and polished wood or "woodgrain-look" dashboards. Compared to mainstream vehicles, luxury cars have traditionally emphasized comfort and safety. Luxury vehicles are also a status symbol for conspicuous consumption.

Contemporary luxury cars also offer higher performance and better handling, which is often known as sport luxury. However, in Europe, where large-displacement engines are often heavily taxed and many luxury buyers shy away from conspicuous consumption, brands offer buyers the option of removing exterior engine-identifying badges.

Forbes noted that the reputation of luxury marques enables them to continually introduce many new safety technologies and comfort amenities, such as anti-lock brakes, electronic stability control and DVD entertainment systems, long before they trickle down to mass market cars. Numerous "smart car" features were found on luxury cars as early as 2009.

Layout and Powertrain

Front mounted longitudinal engines with rear-wheel drive (FR) is a common layout of luxury cars. European marques like Mercedes-Benz, BMW, and Jaguar have almost never adopted front mounted engines with front-wheel drive (FF) and retained a line-up mostly or entirely made up of FR cars. Japanese brands such as Lexus and Infiniti also have predominantly FR lineups. The FR layout, while more expensive than the FF, has been retained by these luxury manufacturers as it allows for higher performance engines (particularly the straight-6, V8, and other engine configurations with more cylinders), better handling, and a smoother ride.

American manufacturers also largely followed the FR for their luxury brands (as well as their mass-market cars of the time). However, due to the Arab Oil Embargo of 1973 and the 1979 fuel crises, began eliminating their FR platforms in favor of the more economical front wheel drive transverse engine layout (FF). Chrysler went 100% FF by 1990 and GM's Cadillac and Buick brands for the US were entirely FF by 1997. One of the few notable holdouts was Ford's Lincoln Town Car and Lincoln LS.

In the 21st century, as part of the revamp of its design and image, Cadillac returned most of its lineup (sedans, roadsters, crossovers and SUVs) to rear- or all-wheel-drive, the only exceptions being the front-wheel drive Cadillac BLS (since discontinued; never sold in North America); Cadillac DTS (also discontinued); the Cadillac XTS (which replaced the DTS); and Cadillac ELR (a plug-in hybrid). Chrysler returned its full-size cars to this layout with the Chrysler 300 using shared technologies during the DaimlerChrysler era (Chrysler and Daimler-Benz (the parent company of Mercedes-Benz) co-developed the LX platform currently retained by Fiat Chrysler Automobiles. Ford's Lincoln division retained the longtime FR platform for the Town Car (phased out in 2011 with no direct replacement), intended for use as a limousine and chauffeured car. However, newer offerings such as the MKZ and MKS use newer FF platforms shared with mainstream Ford vehicles, the Ford Fusion and Ford Taurus, respectively, with all-wheel drive as an option.

History and Sales

Pre-WW2

Rolls-Royce Phantom II Continental was a British luxury car

A wide array of European producers made luxury cars, including Rolls-Royce Limited, Bugatti, Delage, Delahaye, and Talbot-Lago made vehicles that are still collectible today. Similarly, the Americans built the Duesenberg, Packard, Pierce Arrow, Cadillac,

Chrysler, Lincoln and Cord luxury models.

1946-1990

Germany slowly became an export powerhouse during this period, building on success with the Mercedes-Benz brand, later joined by BMW, which now owns Rolls-Royce Motor Cars, and Volkswagen, which now owns Audi, Bentley, and Lamborghini.

In the United States luxury market, Cadillac and Lincoln Motor Company had long been the best-selling and second best-selling luxury brands until 1998, when they were overtaken by Japanese and German brands. These cars were not relevant in most export markets; they were tailored to the U.S. market - moderate price point luxury vehicles not constrained by gasoline pricing.

The U.S. developed a special vehicle category, the personal luxury car, a stylized, mass produced 2 door coupé, relying on standard components. These vehicles were popular from 1958 to 1980.

Before World War II, France was a leading producer of powerful luxury automobiles. After World War II, the French government used *puissance fiscale* tax regulations to encourage manufacturers to build cars with small engines, and French motorists to buy them. Since many export markets do not have this same constraint, the French automobile industry has become less relevant as an export power and currently produces no luxury brands.

Italy focused on high end sports cars in the luxury market, such as the Ferrari and Maserati.

1990-2007

Since the 1980s, a host of new manufacturers have entered the luxury market to challenge the traditional players. The three major Japanese auto manufacturers, Honda, Toyota, and Nissan, created their respective luxury brands particularly for the US. As a result of voluntary export restraints imposed in 1981, these manufacturers were limited to a number of vehicles they could export. While these companies sidestepped this by establishing US production facilities for mass market vehicles, their home factories soon begun producing higher-priced cars as they carried a greater profit margin per car. Acura was launched in 1986, while Lexus and Infiniti were unveiled in 1989. By 1992, these three divisions had sales of over US$3.5 billion, using lower prices and innovation to take market share from both domestic (Cadillac, Lincoln) and the European (Mercedes-Benz, Volvo, BMW, Audi and Jaguar) luxury car makers.

Mercedes-Benz and BMW were the top-selling luxury import until 1991, when they were overtaken by Lexus. Since 2000, Lexus has been the number-one-selling luxury car marque in the U.S.

Since the 2000s, with the Cadillac CTS, the marque has seen a resurgence in sales and brand value. Ford's Lincoln, which had seen sales fall as a result of an aging lineup, has attempted to return that luxury marque to competitiveness, by releasing new models such as the Lincoln MKS, as well as divesting itself of its other Premier Automotive Group brands.

Hyundai had recently released the Genesis and the Equus, hoping to repeat the same strategy of undercutting their established competitors.

2008 Recession

The Late-2000s recession was the first time since the Great Depression of the 1930s that the luxury car market suffered considerably, something not seen in previous economic downturns. Many such customers saw their net worth decline following the collapse in financial markets and real-estate values. For example, some of the steepest dropoffs came at the high end, including the BMW 7 Series and Rolls-Royce Phantom, and Mercedes-Benz unexpectedly dropped the starting price of its all-new 2010 E-Class. The unusually sharp decline in luxury car sales have led observers to believe that there is a fundamental shift and reshaping of the luxury automotive market, with one industry official suggesting that the marques no longer command the premiums that they used to, and another saying that conspicuous consumption was no longer attractive in poor economic conditions. Additionally, mainstream brands have been able to offer amenities and devices such as leather, wood, and anti-lock brakes, previously found only on luxury cars, as the costs decline.

However, luxury vehicle sales have not collapsed as much as their non-luxury counterparts. Luxury vehicle marques generally benefited from financially healthier dealerships, better leasing and certified pre-owned programs and loyal customers, so sales are expected to rebound more quickly than mass market cars. Others note that there is growing interest in luxury vehicles from emerging markets such as China and Russia.

The entry-level luxury segment has been very competitive, and there has been price-overlapping with well-equipped non-luxury cars. For example, in Canada, several luxury manufacturers set sales records in August 2009, due mostly to aggressive incentives on entry-level luxury vehicles. In September 2009, BMW, Mercedes-Benz, Lexus and Audi all saw their Canadian sales increase by more than 10 per cent compared to a year earlier, despite overall Canadian auto sales being down 3.5 per cent compared to September 2008. The head of Mercedes-Benz Canada suggested that the brand "has been able to attract 'middle-class' consumers even during the recession because of the sense that owning a Mercedes-Benz comes with 'membership in a club'." BMW Canada's chief said luxury cars continued to be attractive, "I think due to new product offensives and due to new design and due to the fact that we are the benchmark in all areas when it comes to fuel efficiency... that together stimulates a lot of the market". BMW has managed to remain profitable in 2009 while other competitors were posting losses, by scaling down production quickly to avoid cash burn through bloated inventories.

Recent Trends

In 2010, BMW was the best-selling luxury vehicle manufacturer by sales, ahead of rivals Audi AG and Daimler AG's Mercedes-Benz marque.

References

- Croft, Robin (2006). "Folklore, families and fear: understanding consumption decisions through the oral tradition". Journal of Marketing Management. 22 (9/10): 1053–1076. ISSN 0267-257X doi:10.1362/026725706778935574
- Babaian, Sharon (1998). The Most Benevolent Machine: A Historical Assessment of Cycles in Canada. Ottawa: National Museum of Science and Technology. p. 97. ISBN 0-660-91670-3
- Bargmann, Joe (2013-11-04). "Urbee 2, the 3D-Printed Car That Will Drive Across the Country". Popular Mechanics. Retrieved 2014-02-10
- Bell, Jonathan (2003). Concept Car Design: Driving the Dream. Rotovision. p. 67. ISBN 978-2-88046-564-3. Retrieved 9 January 2014
- Code of Federal Regulations, Title 40, Protection of Environment, PT. 425 699. US: Office of the Federal Register. 2010. p. 862. Retrieved 1 March 2016
- Jain, Sarah S. Lochlann (February 2004). "'Dangerous Instrumentality': The Bystander as Subject in Automobility" (PDF). Cultural Anthropology. 19 (1): 61–94. doi:10.1525/can.2004.19.1.61
- Fuhrmans, Vanessa (2011-08-08). "A Handful of Maybachs Isn't Enough". The Wall Street Journal. Retrieved 2014-03-11
- Sperling, Daniel; Kurani, Ken (September 2001). Transportation, Energy, and Environmental Policy. Transportation Research Board. p. 230. ISBN 0-309-08571-3
- Schoenberger, Robert (2008-06-06). "As buyers shun SUVs, expect to pay more for that small car". Cleveland Business News. Retrieved 2010-11-16
- Rogoski, Richard R. (2008-12-01). "Recession not a roadblock for luxury-car sales". Triangle Business Journal. Retrieved 2014-03-11
- Fuhs, Allen (2008-09-17). Hybrid Vehicles: and the Future of Personal Transportation. CRC Press. p. 7. ISBN 9781420075342
- Foster, Patrick (January 2008). "1963 Jeep Wagoneer - Landmark Vehicle". 4 Wheel Drive & Sport Utility Magazine. Retrieved 2010-11-16

3

Motorcycle Design and Technologies

Motorcycle design is the art of designing and engineering motorcycles. Motorcycle braking systems, motorcycle fork, shock absorber, hub-center steering, sidecar, featherbed frame and motorcycle fairing has been explained. The chapter on motorcycle design and technologies offers an insightful focus, keeping in mind the complex subject matter.

Motorcycle Design

Motorcycle design can be described as activities that define the appearance, function and engineering of motorcycles.

Professionally it is a branch of industrial design, similar to automotive design using identical techniques and methodology, but confined by a set of conventions about what is acceptable to the buying public. These conventions have been defined by the acceptance of the industry and media as a whole to the assumption that the public will only purchase machines that bear more than a passing resemblance to competition machines of whatever kind. In some large OEM motorcycle manufacturers, the term designer can also be applied to the project leader or chief engineer charged with laying down the principal architecture of the vehicle. In recent years it has also become associated with custom or "chopper" builder culture.

Professional Design

Professional motorcycle designers almost always hold degrees in industrial design, industrial design engineering or similar, and have training in styling, modeling, as well as knowledge in aspects of technology associated with single track vehicles. Although no degree as a specialisation exists per se, the majority of candidates graduate through colleges and universities with established transportation design courses, and are trained as automotive designers.

Most OEM motorcycle manufacturers, such as Honda, Suzuki, Kawasaki, BMW, Ducati, Piaggio and others have in-house design studios dedicated to this purpose, while others such as Yamaha and KTM depend on specialised independent design consultancies.

Methodology

Design and Engineering Relationships

Due to the high importance of mechanical components or even exposed engines to motorcycle styling, almost always designers will have a greater sensitivity to and awareness of engineering than will typical car designers. In OEM situations, large teams of professional engineers and specialists will collaborate on each project development, allowing the designer to focus on the more intangible or subjective aspects of design, such as styling, human-machine interface psychology, and market and cultural relationships. In other matters such as pure mechanical ergonomics (such as seat height, handlebar placement, etc.), or basic layout (the location of major components, storage, etc.) there is usually considerable overlap between the designer and engineer. The designer will nominally approach each problem from a human interface, or "feel" or "irrational" point of view (example : "Does this material feel cold or warm, and is this feeling appropriate to this vehicle's target consumer?"), while the engineer will attack each problem with the "rational" or clinical approach of empirically weighing the cause and effect of each design decision against the project's technical and economic design targets (example : "Can this material be moulded into the designer's desired shape? Will that be too expensive to produce?")

Research and Concept Design

In OEM motorcycle design, the normal procedure of developing a new motorcycle involves the same steps in other professional design disciplines : identifying a target consumer, researching them to identify benchmarks and project targets, then proposing concept directions in a written form known as a Design Brief or QFD. From this point, artwork is developed to visually communicate the designer's ideas. These are presented in 2Db drawing or illustrated form, from which a winning direction is down selected for further development. Once a satisfactory design is established on paper (the term paper is a generalization that can include traditional hand renderings, digital artwork or CAD drawings), then full scale modeling begins to realise the design in tangible 3D form.

Styling

Often used as an interchangeable term with "design", styling is in fact just one component of the design process. Typically, styling is developed through sketches, renderings and illustrations then realised in 3D form using automotive styling clay, specialised industrial modeling foams such as Sibatool, Renshape or Epiwood, or in increasingly limited cases plaster or body filler. As the most subjective part of the design process, the various members of the development team must depend heavily on the judgment, skill and experience of the appointed designer to create an appropriate look.

The most misunderstood element and the most dangerous to the success of a product, is the idea that team members should evaluate the design based on personal tastes or preferences. Industrial design is not an art form, but a focused creative expression using the scientific data and analysis in the Design Brief and QFD as ultimate guidelines. The target user, their needs and tastes should be reflected in the final design, not necessarily exclusively those of the design team. Of course, many complex variables such as the OEM brand identity, past successes and failures, and whimsical trends often skew or distort styling decisions. In instances where the factors are overwhelming, OEM's may err on the side of cautious conservative design.

Parallel Development

Because of the need to reduce development time and costs, the "styling" design model is usually developed in parallel with the engineering 3D design. While there is an increasing amount of digital design input in the modern OEM design process, nearly all major motorcycle manufacturers still rely on full scale clay models to render the master style model, then scan and import the styling surfaces into suitable 3D software packages (Alias, CATIA, ISEM Surf) for integration into the 3D engineering CAD platform (CATIA, ProEngineer, etc.). Once combined, the design team can virtually refine the motorcycle by optimising component assembly, checking for any undesirable interferences between parts, and predict and eliminate possible engineering problems. Typically, designers and engineers will have the greatest number of conflicts during this phase of development, as designers will fight to maintain the original styling and design of the clay model and artwork into the production vehicle, while the engineer will eliminate all problems in the most efficient manner possible. The success of the final product depends heavily on the level of cooperation between these often conflicting needs.

Amateur and Specialists

Custom Builders

In recent years, largely due to the popularity of television programs like *Orange County Chopper* and *Biker Build-off*, the building of one of a kind "chopper" or "cruiser" type motorcycles has become more mainstream, leading to a flourishing builder industry. As a whole, these vehicles are not designed in the professional sense, but rather crafted by hand by metal workers and artisans using traditional skills. The resulting vehicles tend to be very elaborate, expensive and difficult or impossible to reproduce in mass production, but are highly valued for the same reasons.

Among custom motorcycle culture, certain names have become famous for their creations and have led to mainstream acceptance of previously unacceptable design solutions such as extreme ergonomics, totally rigid rear wheels without the benefit of suspension, minimal lighting and limited ground clearance for cornering. These design characteristics are purely emotional in nature, being led by styling and image rather than technical or performance considerations.

"Specials"

Custom and specials motorcycles are similar to the above but tend to be super sport type motorcycles, or at least high-performance based, using many special add-on parts, one-of-a-kind or limited series frames, racing wheels and parts or hand-made components to maximise performance. While modifying motorcycles is an activity as old as the motorcycle itself, the "special" culture or "streetfighter" began to flourish in the mid-1970s as a response to the myriad high performance Japanese motorcycles then available, but whose power far exceeded their handling. Individuals would choose pre-manufactured parts from catalogs or from other bikes and redesign their particular machine to suit their desires. In general this activity is limited to one-of-a-kind vehicles and, as with custom motorcycles, uses very little genuine engineering or design methodology, although some small-scale manufacturers exist who make limited runs of a given model. In some cases, these tiny specialists were successful enough to grow into full-scale OEM companies such as the Buell Motorcycle Company and Bimota of Italy.

Bicycle and Motorcycle Dynamics

Bicycle and motorcycle dynamics is the science of the motion of bicycles and motorcycles and their components, due to the forces acting on them. Dynamics falls under a branch of physics known as classical mechanics. Bike motions of interest include balancing, steering, braking, accelerating, suspension activation, and vibration. The study of these motions began in the late 19th century and continues today.

A computer-generated, simplified model of bike and rider demonstrating an uncontrolled right turn.

Bicycles leaning in a turn.

Bicycles and motorcycles are both single-track vehicles and so their motions have many fundamental attributes in common and are fundamentally different from and more difficult to study than other wheeled vehicles such as dicycles, tricycles, and quadracycles. As with unicycles, bikes lack lateral stability when stationary, and under most circumstances can only remain upright when moving forward. Experimentation and mathematical analysis have shown that a bike stays upright when it is steered to keep its center of mass over its wheels. This steering is usually supplied by a rider, or in certain circumstances, by the bike itself. Several factors, including geometry, mass distribution, and gyroscopic effect all contribute in varying degrees to this self-stability, but long-standing hypotheses and claims that any single effect, such as gyroscopic or trail, is solely responsible for the stabilizing force have been discredited.

While remaining upright may be the primary goal of beginning riders, a bike must lean in order to maintain balance in a turn: the higher the speed or smaller the turn radius, the more lean is required. This balances the roll torque about the wheel contact patches generated by centrifugal force due to the turn with that of the gravitational force. This lean is usually produced by a momentary steering in the opposite direction, called countersteering. Countersteering skill is usually acquired by motor learning and executed via procedural memory rather than by conscious thought. Unlike other wheeled vehicles, the primary control input on bikes is steering torque, not position.

Although longitudinally stable when stationary, bikes often have a high enough center of mass and a short enough wheelbase to lift a wheel off the ground under sufficient acceleration or deceleration. When braking, depending on the location of the combined center of mass of the bike and rider with respect to the point where the front wheel contacts the ground, bikes can either skid the front wheel or flip the bike and rider over the front wheel. A similar situation is possible while accelerating, but with respect to the rear wheel.

History

The history of the study of bike dynamics is nearly as old as the bicycle itself. It includes contributions from famous scientists such as Rankine, Appell, and Whipple. In the early 19th century Karl von Drais, credited with inventing the two-wheeled vehicle variously called the laufmaschine, velocipede, draisine, and dandy horse, showed that a rider could balance his device by steering the front wheel. In 1869, Rankine published an article in *The Engineer* repeating von Drais's assertion that balance is maintained by steering in the direction of a lean.

Draisine.

In 1897, the French Academy of Sciences made understanding bicycle dynamics the goal of its Prix Fourneyron competition. Thus, by the end of the 19th century, Carlo Bourlet, Emmanuel Carvallo, and Francis Whipple had showed with rigid-body dynamics that some safety bicycles could actually balance themselves if moving at the right speed. Bourlet won the Prix Fourneyron, and Whipple won the Cambridge University Smith Prize. It is not clear to whom should go the credit for tilting the steering axis from the vertical which helps make this possible.

In 1970, David E. H. Jones published an article in *Physics Today* showing that gyroscopic effects are not necessary to balance a bicycle. Since 1971, when he identified and named the wobble, weave and capsize modes, Robin Sharp has written regularly about the behavior of motorcycles and bicycles. While at Imperial College, London, he worked with David Limebeer and Simos Evangelou.

In the early 1970s, Cornell Aeronautical Laboratory (CAL, later Calspan Corporation in Buffalo, NY USA) was sponsored by the Schwinn Bicycle Company and others to study and simulate bicycle and motorcycle dynamics. Portions of this work have now been released to the public and scans of over 30 detailed reports have been posted at this TU Delft Bicycle Dynamics site.

Since the 1990s, Cossalter, et al., have been researching motorcycle dynamics at the University of Padova. Their research, both experimental and numerical, has covered weave, wobble, chatter, simulators, vehicle modelling, tire modelling, handling, and minimum lap time maneuvering.

In 2007, Meijaard, et al., published the canonical linearized equations of motion, in the *Proceedings of the Royal Society A,* along with verification by two different methods. These equations assumed the tires to roll without slip, that is to say, to go where they point, and the rider to be rigidly attached to the rear frame of the bicycle.

In 2011, Kooijman, et al., published an article in *Science* showing that neither gyroscopic effects nor so-called caster effects due to trail are necessary for a bike to balance itself. They designed a two-mass-skate bicycle that the equations of motion predict is self-stable even with negative trail, the front wheel contacts the ground in front of the steering axis, and with counter-rotating wheels to cancel any gyroscopic effects. Then they constructed a physical model to validate that prediction. This may require some of the details provided below about steering geometry or stability to be re-evaluated. Bicycle dynamics was named 26 of *Discover*'s 100 top stories of 2011.

In 2013, Eddy Merckx Cycles was awarded over €150,000 with Ghent University to examine bicycle stability.

External forces on a bike and rider leaning in a turn: Weight in green, drag in blue, vertical ground reaction in red, net propulsive and rolling resistance in yellow, friction in response to turn in orange, and net torques on front wheel in magenta.

Spring between front fork and rear frame

Forces

If the bike and rider are considered to be a single system, the forces that act on that system and its components can be roughly divided into two groups: internal and external. The external forces are due to gravity, inertia, contact with the ground, and contact with the atmosphere. The internal forces are caused by the rider and by interaction between components.

External Forces

As with all masses, gravity pulls the rider and all the bike components toward the earth. At each tire contact patch there are ground reaction forces with both horizontal and vertical components. The vertical components mostly counteract the force of gravity, but also vary with braking and accelerating. The horizontal components, due to friction between the wheels and the ground, including rolling resistance, are in response to propulsive forces, braking forces, and turning forces. Aerodynamic forces due to the atmosphere are mostly in the form of drag, but can also be from crosswinds. At normal bicycling speeds on level ground, aerodynamic drag is the largest force resisting forward motion. At faster speed, aerodynamic drag becomes overwhelmingly the largest force resisting forward motion.

Turning forces are generated during maneuvers for balancing in addition to just changing direction of travel. These may be interpreted as centrifugal forces in the accelerating reference frame of the bike and rider; or simply as inertia in a stationary, inertial reference frame and not forces at all. *Gyroscopic* forces acting on rotating parts such as wheels, engine, transmission, etc., are also due to the inertia of those rotating parts.

Internal Forces

Internal forces, those between components of the bike and rider system, are mostly caused by the rider or by friction. In addition to pedaling, the rider can apply torques between the steering mechanism (front fork, handlebars, front wheel, etc.) and rear frame, and between the rider and the rear frame. Friction exists between any parts that move against each other: in the drive train, between the steering mechanism and the rear frame, etc. In addition to brakes, which create friction between rotating wheels and non-rotating frame parts, many bikes have front and rear suspensions. Some motorcycles and bicycles have a steering damper to dissipate undesirable kinetic energy, and some bicycles have a spring connecting the front fork to the frame to provide a progressive torque that tends to steer the bicycle straight ahead. On bikes with rear suspensions, feedback between the drive train and the suspension is an issue designers attempt to handle with various linkage configurations and dampers.

Motions

Motions of a bike can be roughly grouped into those out of the central plane of symmetry: lateral; and those in the central plane of symmetry: longitudinal or vertical. Lateral motions include balancing, leaning, steering, and turning. Motions in the central plane of symmetry include rolling forward, of course, but also stoppies, wheelies, brake diving, and most suspension activation. Motions in these two groups are linearly decoupled, that is they do not interact with each other to the first order. An uncontrolled bike is laterally unstable when stationary and can be laterally self-stable when moving under the right conditions or when controlled by a rider. Conversely, a bike is longitudinally stable when stationary and can be longitudinally unstable when undergoing sufficient acceleration or deceleration.

Lateral Dynamics

Of the two, lateral dynamics has proven to be the more complicated, requiring three-dimensional, multibody dynamic analysis with at least two generalized coordinates to analyze. At a minimum, two coupled, second-order differential equations are required to capture the principal motions. Exact solutions are not possible, and numerical methods must be used instead. Competing theories of how bikes balance can still be found in print and online. On the other hand, much longitudinal dynamic analysis can be accomplished simply with planar kinetics and just one coordinate.

Balance

When discussing bike balance, it is necessary to distinguish carefully between "stability", "self-stability", and "controllability". Recent research suggests that "rider-controlled stability of bicycles is indeed related to their self-stability."

Balancing a bicycle by keeping the wheels under the center of mass

A bike remains upright when it is steered so that the ground reaction forces exactly balance all the other internal and external forces it experiences, such as gravitational if leaning, inertial or centrifugal if in a turn, gyroscopic if being steered, and aerodynamic if in a crosswind. Steering may be supplied by a rider or, under certain circumstances, by the bike itself. This self-stability is generated by a combination of several effects that depend on the geometry, mass distribution, and forward speed of the bike. Tires, suspension, steering damping, and frame flex can also influence it, especially in motorcycles.

Even when staying relatively motionless, a rider can balance a bike by the same principle. While performing a track stand, the rider can keep the line between the two contact patches under the combined center of mass by steering the front wheel to one side or the other and then moving forward and backward slightly to move the front contact patch from side to side as necessary. Forward motion can be generated simply by pedaling. Backwards motion can be generated the same way on a fixed-gear bicycle. Otherwise, the rider can take advantage of an opportune slope of the pavement or lurch the upper body backwards while the brakes are momentarily engaged.

If the steering of a bike is locked, it becomes virtually impossible to balance while riding. On the other hand, if the gyroscopic effect of rotating bike wheels is cancelled by adding counter-rotating wheels, it is still easy to balance while riding. One other way that a bike can be balanced, with or without locked steering, is by applying appropriate torques between the bike and rider similar to the way a gymnast can swing up from hanging straight down on uneven parallel bars, a person can start swinging on a swing from rest by pumping their legs, or a double inverted pendulum can be controlled with an actuator only at the elbow.

Forward Speed

The rider applies torque to the handlebars in order to turn the front wheel and so to control lean and maintain balance. At high speeds, small steering angles quickly move the ground contact points laterally; at low speeds, larger steering angles are required to achieve the same results in the same amount of time. Because of this, it is usually easier to maintain balance at high speeds. Also, self-stability usually only occurs at speeds above some minimum, and so going faster increases the chances that a bike is contributing to its own stability.

Center of Mass Location

The farther forward (closer to front wheel) the center of mass of the combined bike and rider, the less the front wheel has to move laterally in order to maintain balance. Conversely, the farther back (closer to the rear wheel) the center of mass is located, the more front wheel lateral movement or bike forward motion is required to regain balance. This can be noticeable on long-wheelbase recumbents, choppers, and wheelie bikes. It can also be a challenge for touring bikes that carry a heavy load of gear over or even behind the rear wheel. Mass over the rear wheel can be more easily controlled if it is lower than mass over the front wheel.

A bike is also an example of an inverted pendulum. Just as a broomstick is more easily balanced in the hand than a pencil, a tall bike (with a high center of mass) can be easier to balance when ridden than a low one because the tall bike's lean rate (rate at which its angle of lean increases as it begins to fall over) will be slower. However, a rider can have the opposite impression of a bike when it is stationary. A top-heavy bike can require more effort to keep upright, when stopped in traffic for example, than a bike which is just as tall but with a lower center of mass. This is an example of a vertical second-class lever. A small force at the end of the lever, the seat or handlebars at the top of the bike, more easily moves a large mass if the mass is closer to the fulcrum, where the tires touch the ground. This is why touring cyclists are advised to carry loads low on a bike, and panniers hang down on either side of front and rear racks.

Trail

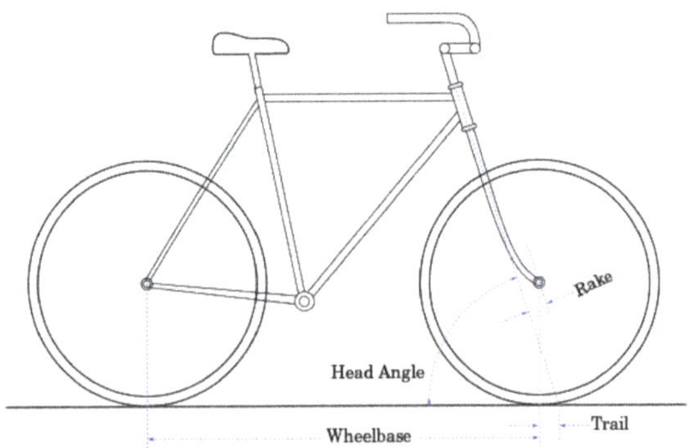

Bicycle head angle, rake, and trail

A factor that influences how easy or difficult a bike will be to ride is trail, the distance that the front wheel ground contact point trails behind the steering axis ground contact point. The steering axis is the axis about which the entire steering mechanism (fork, handlebars, front wheel, etc.) pivots. In traditional bike designs, with a steering axis tilted back from the vertical, positive trail tends to steer the front wheel into the direction of a lean, independent of forward speed. This can be simulated by pushing a stationary bike to one side. The front wheel will usually also steer to that side. In a lean, gravity provides this force. The dynamics of a moving bike are more complicated, however, and other factors can contribute to or detract from this effect.

Trail is a function of head angle, fork offset or rake, and wheel size. Their relationship can be described by this formula:

$$\text{Trail} = \frac{(R_w \cos(A_h) - O_f)}{\sin(A_h)}$$

where R_w is wheel radius, A_h is the head angle measured clock-wise from the horizontal and O_f is the fork offset or rake. Trail can be increased by increasing the wheel size, decreasing the head angle, or decreasing the fork rake.

The more trail a traditional bike has, the more stable it feels, although too much trail can make a bike feel difficult to steer. Bikes with negative trail (where the contact patch is in front of where the steering axis intersects the ground), while still rideable, are reported to feel very unstable. Normally, road racing bicycles have more trail than touring bikes but less than mountain bikes. Mountain bikes are designed with reduced head angles than road bikes to improve stability for descents, and therefore have greater trail. Touring bikes are built with small trail to allow the rider to control a bike weighed down with baggage. As a consequence, an unloaded touring bike can feel unstable. In bicycles, fork rake, often a curve in the fork blades forward of the steering axis, is used to diminish trail. Bikes with negative trail exist, such as the Python Lowracer, and are rideable, and an experimental bike with negative trail has been shown to be self-stable.

In motorcycles, rake refers to the head angle instead, and offset created by the triple tree is used to diminish trail.

A small survey by Whitt and Wilson found:

- touring bicycles with head angles between 72° and 73° and trail between 43 mm and 60 mm
- racing bicycles with head angles between 73° and 74° and trail between 28 mm and 45 mm
- track bicycles with head angles of 75° and trail between 23.5 mm and 37 mm.

However, these ranges are not hard and fast. For example, LeMond Racing Cycles offers both with forks that have 45 mm of offset or rake and the same size wheels:

- a 2006 Tete de Course, designed for road racing, with a head angle that varies from 71¼° to 74°, depending on frame size, and thus trail that varies from 51.5 mm to 69 mm.
- a 2007 Filmore, designed for the track, with a head angle that varies from 72½° to 74°, depending on frame size, and thus trail that varies from 51.5 mm to 61 mm.

The amount of trail a particular bike has may vary with time for several reasons. On bikes with front suspension, especially telescopic forks, compressing the front suspension, due to heavy braking for example, can steepen the steering axis angle and reduce trail. Trail also varies with lean angle, and steering angle, usually decreasing from a maximum when the bike is straight upright and steered straight ahead. Trail can decrease to zero with sufficiently large lean and steer angles, which can alter how stable a bike feels. Finally, even the profile of the front tire can influence how trail varies as the bike is leaned and steered.

A measurement similar to trail, called either mechanical trail, normal trail, or true trail, is the *perpendicular* distance from the steering axis to the centroid of the front wheel contact patch.

Wheelbase

A factor that influences the directional stability of a bike is wheelbase, the horizontal distance between the ground contact points of the front and rear wheels. For a given displacement of the front wheel, due to some disturbance, the angle of the resulting path from the original is inversely proportional to wheelbase. Also, the radius of curvature for a given steer angle and lean angle is proportional to the wheelbase. Finally, the wheelbase increases when the bike is leaned and steered. In the extreme, when the lean angle is 90°, and the bike is steered in the direction of that lean, the wheelbase is increased by the radius of the front and rear wheels.

Steering Mechanism Mass Distribution

Another factor that can also contribute to the self-stability of traditional bike designs is the distribution of mass in the steering mechanism, which includes the front wheel, the fork, and the handlebar. If the center of mass for the steering mechanism is in front of the steering axis, then the pull of gravity will also cause the front wheel to steer in the direction of a lean. This can be seen by leaning a stationary bike to one side. The front wheel will usually also steer to that side independent of any interaction with the ground. Additional parameters, such as the fore-to-aft position of the center of mass and the elevation of the center of mass also contribute to the dynamic behavior of a bike.

Gyroscopic effects

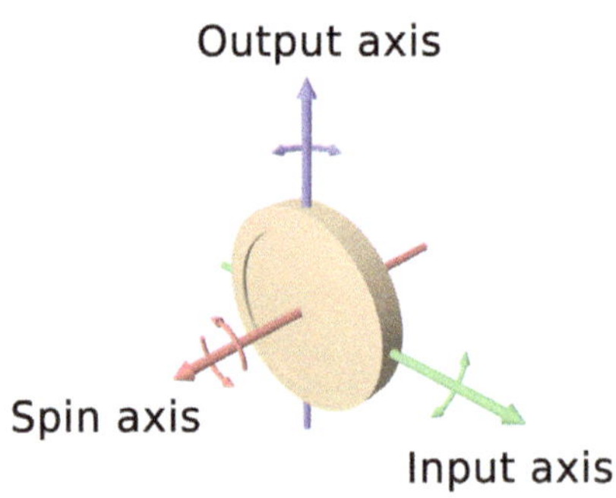

Gyroscopic effect on front wheel of a bike. Applying a torque (in green) about the lean axis results in a reaction torque (in blue) about the steer axis.

The role of the gyroscopic effect in most bike designs is to help steer the front wheel into the direction of a lean. This phenomenon is called *precession* and the rate at which an object precesses is inversely proportional to its rate of spin. The slower a front wheel spins, the faster it will precess when the bike leans, and vice versa. The rear wheel is prevented from precessing as the front wheel does by friction of the tires on the ground, and so continues to lean as though it were not spinning at all. Hence gyroscopic forces do not provide any resistance to tipping.

At low forward speeds, the precession of the front wheel is too quick, contributing to an uncontrolled bike's tendency to oversteer, start to lean the other way and eventually oscillate and fall over. At high forward speeds, the precession is usually too slow, contributing to an uncontrolled bike's tendency to understeer and eventually fall over without ever having reached the upright position. This instability is very slow, on the order of seconds, and is easy for most riders to counteract. Thus a fast bike may feel stable even though it is actually not self-stable and would fall over if it were uncontrolled.

Another contribution of gyroscopic effects is a roll moment generated by the front wheel during countersteering. For example, steering left causes a moment to the right. The moment is small compared to the moment generated by the out-tracking front wheel, but begins as soon as the rider applies torque to the handlebars and so can be helpful in motorcycle racing.

Self-stability

Between the two unstable regimes and influenced by all the factors that contribute to balance (trail, mass distribution, gyroscopic effects, etc.), there may be a range of forward speeds for a given bike design at which these effects steer an uncontrolled bike upright. It has been proven that neither gyroscopic effects nor positive trail are sufficient by themselves or necessary for self-stability, although they certainly can enhance hands-free control.

However, even without self-stability a bike may be ridden by steering it to keep it over its wheels. Note that the effects mentioned above that would combine to produce self-stability may be overwhelmed by additional factors such as headset friction and stiff control cables.

Longitudinal Acceleration

Longitudinal acceleration has been shown to have a large and complex effect on lateral dynamics. In one study, positive acceleration eliminates self stability, and negative acceleration (deceleration) changes the speeds of self stability.

Turning

In order for a bike to turn, that is, change its direction of forward travel, the front wheel must aim approximately in the desired direction, as with any front-wheel steered vehicle. Friction between the wheels and the ground then generates the centripetal acceleration necessary to alter the course from straight ahead as a combination of cornering force and camber thrust. The radius of the turn of an upright (not leaning) bike can be roughly approximated, for small steering angles, by:

$$r = \frac{w}{\delta \cos(\phi)}$$

where r is the approximate radius, w is the wheelbase, δ is the steer angle, and θ is the caster angle of the steering axis.

A Grand Prix motorcyclist leaning in a turn

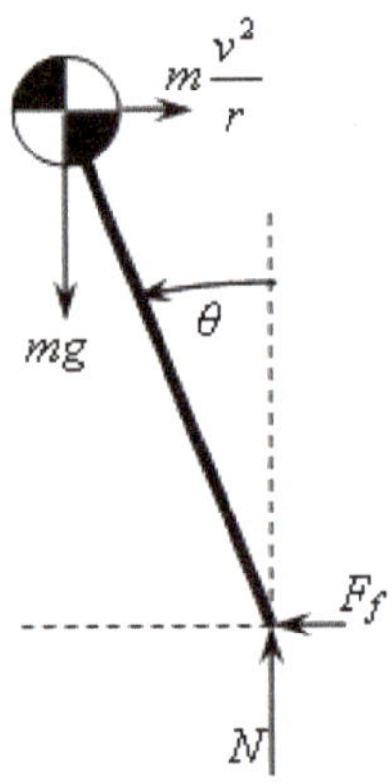

The forces, both physical and inertial, acting on a leaning bike in the rotating reference frame of a turn where N is the normal force, F_f is friction, m is mass, r is turn radius, v is forward speed, and g is the acceleration of gravity.

Leaning

However, unlike other wheeled vehicles, bikes must also lean during a turn to balance the relevant forces: gravitational, inertial, frictional, and ground support. The angle of lean, θ, can easily be calculated using the laws of circular motion:

$$\theta = \arctan\left(\frac{v^2}{gr}\right)$$

where v is the forward speed, r is the radius of the turn and g is the acceleration of gravity. This is in the idealized case. A slight increase in the lean angle may be required on motorcycles to compensate for the width of modern tires at the same forward speed and turn radius.

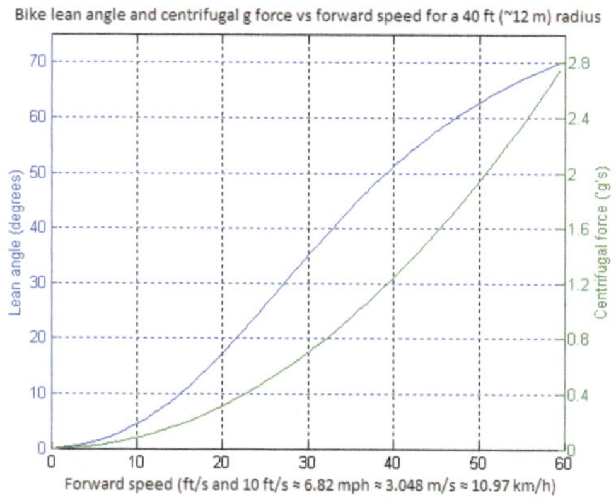

Graph of bike lean angle vs forward speed, assuming unlimited friction between tires and ground.

It can also be seen however that this simple 2-dimensional model, essentially an inverted pendulum on a turntable, predicts that the steady-state turn is unstable. If the bike is displaced slightly downwards from its equilibrium lean angle, the torque of gravity increases, that of centrifugal force decreases and the displacement gets amplified. A more-sophisticated model that allows a wheel to steer, adjust the path, and counter the torque of gravity, is necessary to capture the self-stability observed in real bikes.

For example, a bike in a 10 m (33 ft) radius steady-state turn at 10 m/s (36 km/h, 22 mph) must be at an angle of 45.6°. A rider can lean with respect to the bike in order to keep either the torso or the bike more or less upright if desired. The angle that matters is the one between the horizontal plane and the plane defined by the tire contacts and the location of the center of mass of bike and rider.

This lean of the bike decreases the actual radius of the turn proportionally to the cosine of the lean angle. The resulting radius can be roughly approximated (within 2% of exact value) by:

$$r = \frac{w\cos(\theta)}{\delta\cos(\phi)}$$

where r is the approximate radius, w is the wheelbase, θ is the lean angle, δ is the steer angle, and φ is the caster angle of the steering axis. As a bike leans, the tires' contact patches move farther to the side causing wear. The portions at either edge of a motorcycle tire that remain unworn by leaning into turns is sometimes referred to as chicken strips.

The finite width of the tires alters the actual lean angle of the rear frame from the ideal lean angle described above. The actual lean angle between the frame and the vertical must increase with tire width and decrease with center of mass height. Bikes with fat tires and low center of mass must lean more than bikes with skinnier tires or higher centers of mass to negotiate the same turn at the same speed.

The increase in lean angle due to a tire thickness of $2t$ can be calculated as:

$$\arcsin\left(t\frac{\sin(\phi)}{h-t}\right)$$

where φ is the ideal lean angle, and h is the height of the center of mass. For example, a motorcycle with a 12 inch wide rear tire will have t = 6 inches. If the combined bike and rider center of mass is at a height of 26 inches, then a 25° lean must be increased by 7.28°: a nearly 30% increase. If the tires are only 6 inches wide, then the lean angle increase is only 3.16°, just under half.

It has been shown that the couple created by gravity and the ground reaction forces is necessary for a bicycle to turn at all. On a custom built bicycle with spring-loaded outriggers that exactly cancel this couple, so that the bicycle and rider may assume any lean angle when traveling in a straight line, riders find it impossible to make a turn. As soon as the wheels deviate from a straight path, the bicycle and rider begin to lean in the opposite direction, and the only way to right them is to steer back onto the straight path.

Countersteering

Cyclist riding with no hands on the handlebars.

In order to initiate a turn and the necessary lean in the direction of that turn, a bike must momentarily steer in the opposite direction. This is often referred to as countersteering. With the front wheel now at a finite angle to the direction of motion, a lateral force is developed at the contact patch of the tire. This force creates a torque around the longitudinal (roll) axis of the bike, and this torque causes the bike to lean away from the initially steered direction and toward the direction of the desired turn. Where there is no external influence, such as an opportune side wind to create the force necessary to lean the bike, countersteering is necessary to initiate a rapid turn.

While the initial steer torque and steer angle are both opposite the desired turn direction, this may not be the case to maintain a steady-state turn. The sustained steer angle is usually in the same direction as the turn, but may remain opposite to the direction of the turn, especially at high speeds. The sustained steer torque required to maintain that steer angle is usually opposite the turn direction. The actual magnitude and orientation of both the sustained steer angle and sustained steer torque of a particular bike in a particular turn depend on forward speed, bike geometry, tire properties, and combined bike and rider mass distribution. Once in a turn, the radius can only be changed with an appropriate change in lean angle, and this can be accomplished by additional countersteering out of the turn to increase lean and decrease radius, then into the turn to decrease lean and increase radius. To exit the turn, the bike must again countersteer, momentarily steering more into the turn in order to decrease the radius, thus increasing inertial forces, and thereby decreasing the angle of lean.

Steady-state Turning

Once a turn is established, the torque that must be applied to the steering mechanism in order to maintain a constant radius at a constant forward speed depends on the forward speed and the geometry and mass distribution of the bike. At speeds below the capsize speed, the self-stability of the bike will cause it to tend to steer into the turn, righting itself and exiting the turn, unless a torque is applied in the opposite direction of the turn. At speeds above the capsize speed, the capsize instability will cause it to tend to steer out of the turn, increasing the lean, unless a torque is applied in the direction of the turn. At the capsize speed no input steering torque is necessary to maintain the steady-state turn.

Steering Angle

Several effects influence the steering angle, the angle at which the front assembly is rotated about the steering axis, necessary to maintain a steady-state turn. Some of these are unique to single-track vehicles, while others are also experienced by automobiles.

First, the actual kinematic steering angle, the angle projected onto the road plane to which the front assembly is rotated is a function of the steering angle and the steering axis angle:

$$\Delta = \delta \cos(\phi)$$

where Δ is the kinematic steering angle, δ is the steering angle, and ϕ is the caster angle of the steering axis.

Second, the lean of the bike decreases the actual radius of the turn proportionally to the cosine of the lean angle. The resulting radius can be roughly approximated (within 2% of exact value) by:

$$r = \frac{w\cos(\theta)}{\delta\cos(\phi)}$$

where r is the approximate radius, w is the wheelbase, θ is the lean angle, δ is the steering angle, and ϕ is the caster angle of the steering axis.

Third, because the front and rear tires can have different slip angles due to weight distribution, tire properties, etc., bikes can experience understeer or oversteer. When understeering, the steering angle must be greater, and when oversteering, the steering angle must be less than it would be if the slip angles were equal to maintain a given turn radius. Some authors even use the term *counter-steering* to refer to the need on some bikes under some conditions to steer in the opposite direction of the turn (negative steering angle) to maintain control in response to significant rear wheel slippage.

Fourth, camber thrust contributes to the centripetal force necessary to cause the bike to deviate from a straight path, along with cornering force due to the slip angle, and can be the largest contributor. Camber thrust contributes to the ability of bikes to negotiate a turn with the same radius as automobiles but with a smaller steering angle. When a bike is steered and leaned in the same direction, the camber angle of the front tire is greater than that of the rear and so can generate more camber thrust, all else being equal.

No Hands

While countersteering is usually initiated by applying torque directly to the handlebars, on lighter vehicles such as bicycles, it can also be accomplished by shifting the rider's weight. If the rider leans to the right relative to the bike, the bike leans to the left to conserve angular momentum, and the combined center of mass remains nearly in the same vertical plane. This leftward lean of the bike, called counter lean by some authors, will cause it to steer to the left and initiate a right-hand turn as if the rider had countersteered to the left by applying a torque directly to the handlebars. This technique may be complicated by additional factors such as headset friction and stiff control cables.

The combined center of mass does move slightly to the left when the rider leans to the right relative to the bike, and the bike leans to the left in response. The action, in space, would have the tires move right, but this is prevented by friction between the tires and the ground, and thus pushes the combined center of mass left. This is a small effect, however, as evidenced by the difficulty most people have in balancing a bike by this method alone.

Gyroscopic Effects

As mentioned above in the section on balance, one effect of turning the front wheel is a roll moment caused by gyroscopic precession. The magnitude of this moment is proportional to the moment of inertia of the front wheel, its spin rate (forward motion), the rate that the rider turns the front wheel by applying a torque to the handlebars, and the cosine of the angle between the steering axis and the vertical.

For a sample motorcycle moving at 22 m/s (50 mph) that has a front wheel with a moment of inertia of 0.6 kg·m², turning the front wheel one degree in half a second generates a roll moment of 3.5 N·m. In comparison, the lateral force on the front tire as it tracks out from under the motorcycle reaches a maximum of 50 N. This, acting on the 0.6 m (2 ft) height of the center of mass, generates a roll moment of 30 N·m.

While the moment from gyroscopic forces is only 12% of this, it can play a significant part because it begins to act as soon as the rider applies the torque, instead of building up more slowly as the wheel out-tracks. This can be especially helpful in motorcycle racing.

Two-wheel Steering

Because of theoretical benefits, such as a tighter turning radius at low speed, attempts have been made to construct motorcycles with two-wheel steering. One working prototype by Ian Drysdale in Australia is reported to "work very well." Issues in the design include whether to provide active control of the rear wheel or let it swing freely. In the case of active control, the control algorithm needs to decide between steering with or in the opposite direction of the front wheel, when, and how much. One implementation of two-wheel steering, the Sideways bike, lets the rider control the steering of both wheels directly. Another, the Swing Bike, had the second steering axis in front of the seat so that it could also be controlled by the handlebars.

Milton W. Raymond built a long low two-wheel steering bicycle, called "X-2", with various steering mechanisms to control the two wheels independently. Steering motions included "balance", in which both wheels move together to steer the tire contacts under the center of mass; and "true circle", in which the wheels steer equally in opposite directions and thus steering the bicycle without substantially changing the lateral position of the tire contacts relative to the center of mass. X-2 was also able to go "crabwise" with the wheels parallel but out of line with the frame, for instance with the front wheel near the roadway center line and rear wheel near the curb. "Balance" steering allowed easy balancing despite long wheelbase and low center of mass, but no self-balancing ("no hands") configuration was discovered. True circle, as expected, was essentially impossible to balance, as steering does not correct for misalignment of the tire patch and center of mass. Crabwise cycling at angles tested up to about 45° did not show a tendency to fall over, even under braking. X-2 is mentioned in passing in Whitt and

Wilson's *Bicycling Science* 2nd edition.

Rear-wheel Steering

Because of the theoretical benefits, especially a simplified front-wheel drive mechanism, attempts have been made to construct a rideable rear-wheel steering bike. The Bendix Company built a rear-wheel steering bicycle, and the U.S. Department of Transportation commissioned the construction of a rear-wheel steering motorcycle: both proved to be unrideable. Rainbow Trainers, Inc. in Alton, Illinois, offered US$5,000 to the first person "who can successfully ride the rear-steered bicycle, Rear Steered Bicycle I". One documented example of someone successfully riding a rear-wheel steering bicycle is that of L. H. Laiterman at Massachusetts Institute of Technology, on a specially designed recumbent bike. The difficulty is that turning left, accomplished by turning the rear wheel to the right, initially moves the center of mass to the right, and vice versa. This complicates the task of compensating for leans induced by the environment. Examination of the eigenvalues for bicycles with common geometries and mass distributions shows that when moving in reverse, so as to have rear-wheel steering, they are inherently unstable. Other, purpose-built designs have been published, however, that do not suffer this problem.

Center Steering

Flevobike with center steering

Between the extremes of bicycles with classical front-wheel steering and those with strictly rear-wheel steering is a class of bikes with a pivot point somewhere between the two, referred to as center-steering, and similar to articulated steering. An early implementation of the concept was the Phantom bicycle in the early 1870s promoted as a safer alternative to the penny-farthing. This design allows for simple front-wheel drive and current implementations appear to be quite stable, even rideable no-hands, as many photographs illustrate.

These designs, such as the Python Lowracer, a recumbent, usually have very lax head angles (40° to 65°) and positive or even negative trail. The builder of a bike with negative trail states that steering the bike from straight ahead forces the seat (and thus the rider) to rise slightly and this offsets the destabilizing effect of the negative trail.

Reverse Steering

Bicycles have been constructed, for investigation and demonstration purposes, with the steering reversed so that turning the handlebars to the left causes the front wheel to turn to the right, and vica versa. It is possible to ride such a bicycle, but it has been found that riders experienced with normal bicycles find it very difficult to learn, if they can manage it at all.

Tiller Effect

Tiller effect is the expression used to describe how handlebars that extend far behind the steering axis (head tube) act like a tiller on a boat, in that one moves the bars to the right in order to turn the front wheel to the left, and vice versa. This situation is commonly found on cruiser bicycles, some recumbents, and some motorcycles. It can be troublesome when it limits the ability to steer because of interference or the limits of arm reach.

Tires

Tires have a large influence over bike handling, especially on motorcycles, but also on bicycles. Tires influence bike dynamics in two distinct ways: finite crown radius and force generation. Increase the crown radius of the front tire has been shown to decrease the size or eliminate self stability. Increasing the crown radius of the rear tire has the opposite effect, but to a lesser degree.

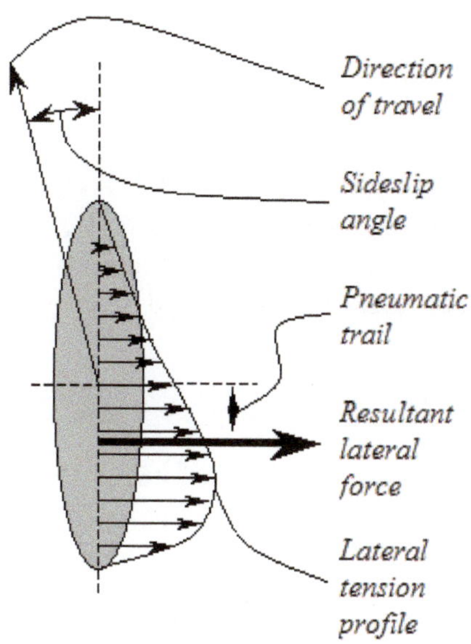

Bike tire contact patch during a right-hand turn

Tires generate the lateral forces necessary for steering and balance through a combination of cornering force and camber thrust. Tire inflation pressures have also been found to be important variables in the behavior of a motorcycle at high speeds. Because the front and rear tires can have different slip angles due to weight distribution, tire properties, etc., bikes can experience understeer or oversteer. Of the two, understeer, in which the front wheel slides more than the rear wheel, is more dangerous since front wheel steering is critical for maintaining balance. Also, because real tires have a finite contact patch with the road surface that can generate a scrub torque, and when in a turn, can experience some side slipping as they roll, they can generate torques about an axis normal to the plane of the contact patch.

One torque generated by a tire, called the self aligning torque, is caused by asymmetries in the side-slip along the length of the contact patch. The resultant force of this side-slip occurs behind the geometric center of the contact patch, a distance described as the pneumatic trail, and so creates a torque on the tire. Since the direction of the side-slip is towards the outside of the turn, the force on the tire is towards the center of the turn. Therefore, this torque tends to turn the front wheel in the direction of the side-slip, away from the direction of the turn, and therefore tends to *increase* the radius of the turn.

Another torque is produced by the finite width of the contact patch and the lean of the tire in a turn. The portion of the contact patch towards the outside of the turn is actually moving rearward, with respect to the wheel's hub, faster than the rest of the contact patch, because of its greater radius from the hub. By the same reasoning, the inner portion is moving rearward more slowly. So the outer and inner portions of the contact patch slip on the pavement in opposite directions, generating a torque that tends to turn the front wheel in the direction of the turn, and therefore tends to *decrease* the turn radius.

The combination of these two opposite torques creates a resulting yaw torque on the front wheel, and its direction is a function of the side-slip angle of the tire, the angle between the actual path of the tire and the direction it is pointing, and the camber angle of the tire (the angle that the tire leans from the vertical). The result of this torque is often the suppression of the inversion speed predicted by rigid wheel models described above in the section on steady-state turning.

High Side

A highsider, highside, or high side is a type of bike motion which is caused by a rear wheel gaining traction when it is not facing in the direction of travel, usually after slipping sideways in a curve. This can occur under heavy braking, acceleration, a varying road surface, or suspension activation, especially due to interaction with the drive train. It can take the form of a single slip-then-flip or a series of violent oscillations.

Maneuverability and Handling

Bike maneuverability and handling is difficult to quantify for several reasons. The geometry of a bike, especially the steering axis angle makes kinematic analysis complicated. Under many conditions, bikes are inherently unstable and must always be under rider control. Finally, the rider's skill has a large influence on the bike's performance in any maneuver. Bike designs tend to consist of a trade-off between maneuverability and stability.

Rider Control Inputs

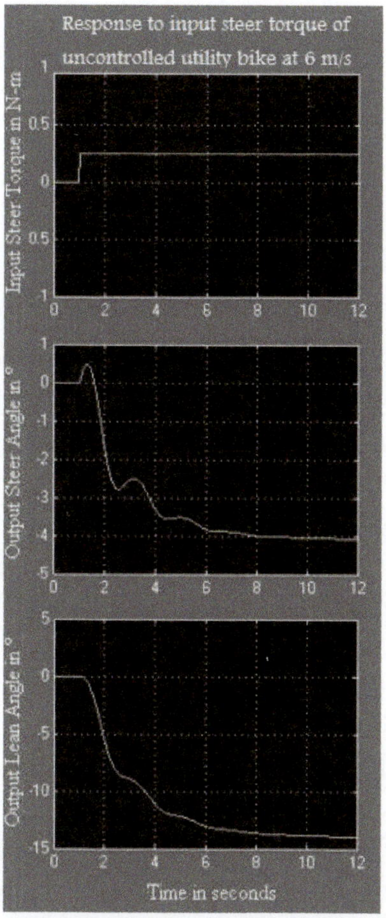

Graphs showing the lean and steer angle response of an otherwise uncontrolled bike, traveling at a forward speed in its stable range (6 m/s), to a steer torque that begins as an impulse and then remains constant. Torque to right causes initial steer to right, lean to left, and eventually a steady-state steer, lean, and turn to left.

The primary control input that the rider can make is to apply a torque directly to the steering mechanism via the handlebars. Because of the bike's own dynamics, due to steering geometry and gyroscopic effects, direct position control over steering angle has been found to be problematic.

A secondary control input that the rider can make is to lean the upper torso relative to the bike. As mentioned above, the effectiveness of rider lean varies inversely with the mass of the bike. On heavy bikes, such as motorcycles, rider lean mostly alters the ground clearance requirements in a turn, improves the view of the road, and improves the bike system dynamics in a very low-frequency passive manner. In motorcycle racing, leaning the torso, moving the body, and projecting a knee to the inside of the turn relative to the bike can also *cause an aerodynamic yawing moment that facilitates entering and rounding the turn.*

Differences from Automobiles

The need to keep a bike upright to avoid injury to the rider and damage to the vehicle even limits the type of maneuverability testing that is commonly performed. For example, while automobile enthusiast publications often perform and quote skidpad results, motorcycle publications do not. The need to “set up” for a turn, lean the bike to the appropriate angle, means that the rider must see further ahead than is necessary for a typical car at the same speed, and this need increases more than in proportion to the speed.

Rating Schemes

Several schemes have been devised to rate the handling of bikes, particularly motorcycles.

- The roll index is the ratio between steering torque and roll or lean angle.
- The acceleration index is the ratio between steering torque and lateral or centripetal acceleration.
- The steering ratio is the ratio between the theoretical turning radius based on ideal tire behavior and the actual turning radius. Values less than one, where the front wheel side slip is greater than the rear wheel side slip, are described as under-steering; equal to one as neutral steering; and greater than one as over-steering. Values less than zero, in which the front wheel must be turned opposite the direction of the curve due to much greater rear wheel side slip than front wheel have been described as counter-steering. Riders tend to prefer neutral or slight over-steering. Car drivers tend to prefer under-steering.
- The Koch index is the ratio between peak steering torque and the product of peak lean rate and forward speed. Large, touring motorcycles tend to have a high Koch index, sport motorcycles tend to have a medium Koch index, and scooters tend to have a low Koch index. It is easier to maneuver light scooters than heavy motorcycles.

Lateral Motion Theory

Although its equations of motion can be linearized, a bike is a nonlinear system. The variable(s) to be solved for cannot be written as a linear sum of independent components, i.e. its behavior is not expressible as a sum of the behaviors of its descriptors. Generally, nonlinear systems are difficult to solve and are much less understandable than linear systems. In the idealized case, in which friction and any flexing is ignored, a bike is a conservative system. Damping, however, can still be demonstrated: under the right circumstances, side-to-side oscillations will decrease with time. Energy added with a sideways jolt to a bike running straight and upright (demonstrating self-stability) is converted into increased forward speed, not lost, as the oscillations die out.

A bike is a nonholonomic system because its outcome is path-dependent. In order to know its exact configuration, especially location, it is necessary to know not only the configuration of its parts, but also their histories: how they have moved over time. This complicates mathematical analysis. Finally, in the language of control theory, a bike exhibits non-minimum phase behavior. It turns in the direction opposite of how it is initially steered.

Degrees of Freedom

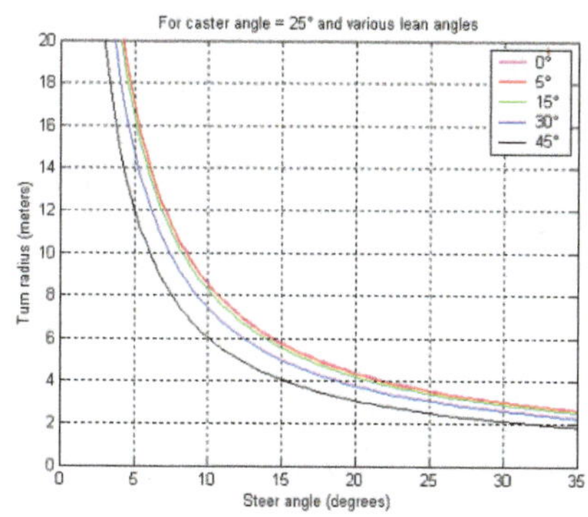

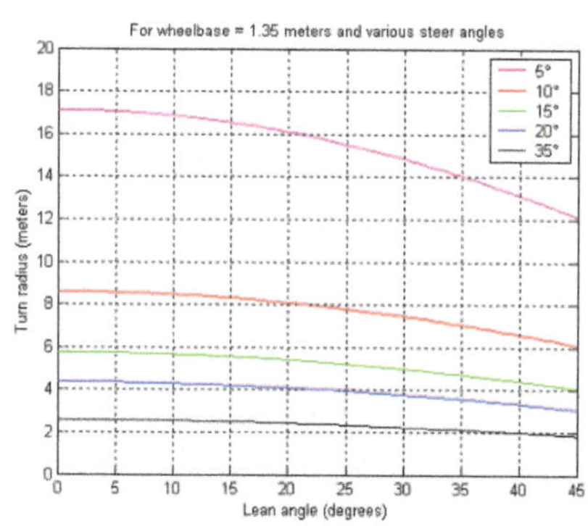

Graphs of bike steer angle and lean angle vs turn radius.

The number of degrees of freedom of a bike depends on the particular model being used. The simplest model that captures the key dynamic features, called the "Whipple model" after Francis Whipple who first developed the equations for it, has four rigid bodies with knife edge wheels rolling without slip on a flat smooth surface, and has 7 degrees of freedom (configuration variables required to completely describe the location and orientation of all 4 bodies):

1. x coordinate of rear wheel contact point
2. y coordinate of rear wheel contact point
3. orientation angle of rear frame (yaw)
4. rotation angle of rear wheel
5. rotation angle of front wheel
6. lean angle of rear frame (roll)
7. steering angle between rear frame and front end

Adding complexity to the model, such as rider movement, suspension movement, tire compliance, or frame flex, adds degrees of freedom. While the rear frame does pitch with leaning and steering, the pitch angle is completely constrained by the requirement for both wheels to remain on the ground, and so can be calculated geometrically from the other seven variables. If the location of the bike and the rotation of the wheels are ignored, the first five degrees of freedom can also be ignored, and the bike can be described by just two variables: lean angle and steer angle.

Equations of Motion

The equations of motion of an idealized bike, consisting of:

- a rigid frame,
- a rigid fork,
- two knife-edged, rigid wheels,
- all connected with frictionless bearings and rolling without friction or slip on a smooth horizontal surface and
- operating at or near the upright and straight-ahead, unstable equilibrium

can be represented by a single fourth-order linearized ordinary differential equation or two coupled second-order differential equations, the lean equation

$$M_{\theta\theta}\ddot{\theta}_r + K_{\theta\theta}\theta_r + M_{\theta\psi}\ddot{\psi} + C_{\theta\psi}\dot{\psi} + K_{\theta\psi}\psi = M_\theta$$

and the steer equation

$$M_{\psi\psi}\ddot{\psi} + C_{\psi\psi}\dot{\psi} + K_{\psi\psi}\psi + M_{\psi\theta}\ddot{\theta}_r + C_{\psi\theta}\dot{\theta}_r + K_{\psi\theta}\theta_r = M_\psi,$$

where

- θ_r is the lean angle of the rear assembly,
- ψ is the steer angle of the front assembly relative to the rear assembly and
- M_θ and M_ψ are the moments (torques) applied at the rear assembly and the steering axis, respectively. For the analysis of an uncontrolled bike, both are taken to be zero.

These can be represented in matrix form as:

$$M\ddot{\mathbf{q}} + C\dot{\mathbf{q}} + K\mathbf{q} = \mathbf{f}$$

where

- M is the symmetrical mass matrix which contains terms that include only the mass and geometry of the bike,
- C is the so-called damping matrix, even though an idealized bike has no dissipation, which contains terms that include the forward speed v and is asymmetric,
- K is the so-called stiffness matrix which contains terms that include the gravitational constant g and v^2 and is symmetric in g and asymmetric in v^2,
- $\mathbf{q}$ is a vector of lean angle and steer angle, and
- $\mathbf{f}$ is a vector of external forces, the moments mentioned above.

In this idealized and linearized model, there are many geometric parameters (wheelbase, head angle, mass of each body, wheel radius, etc.), but only four significant variables: lean angle, lean rate, steer angle, and steer rate. These equations have been verified by comparison with multiple numeric models derived completely independently.

The equations show that the bicycle is like an inverted pendulum with the lateral position of its support controlled by terms representing roll acceleration, roll velocity and roll displacement to steering torque feedback. The roll acceleration term is normally of the wrong sign for self-stabilization and can be expected to be important mainly in respect of wobble oscillations. The roll velocity feedback is of the correct sign, is gyroscopic in nature, being proportional to speed, and is dominated by the front wheel contribution. The roll displacement term is the most important one and is mainly controlled by trail, steering rake and the offset of the front frame mass center from the steering axis. All the terms involve complex combinations of bicycle design parameters and sometimes the speed. The limitations of the benchmark bicycle are considered and extensions to the treatments of tires, frames and riders , and their implications, are included. Optimal rider controls for stabilization and path-following control are also discussed.

Eigenvalues

It is possible to calculate eigenvalues, one for each of the four state variables (lean angle, lean rate, steer angle, and steer rate), from the linearized equations in order to analyze the normal modes and self-stability of a particular bike design. In the plot to the right, eigenvalues of one particular bicycle are calculated for forward speeds of 0–10 m/s (22 mph). When the real parts of all eigenvalues (shown in dark blue) are negative, the bike is self-stable. When the imaginary parts of any eigenvalues (shown in cyan) are non-zero, the bike exhibits oscillation. The eigenvalues are point symmetric about the origin and so any bike design with a self-stable region in forward speeds will not be self-stable going backwards at the same speed.

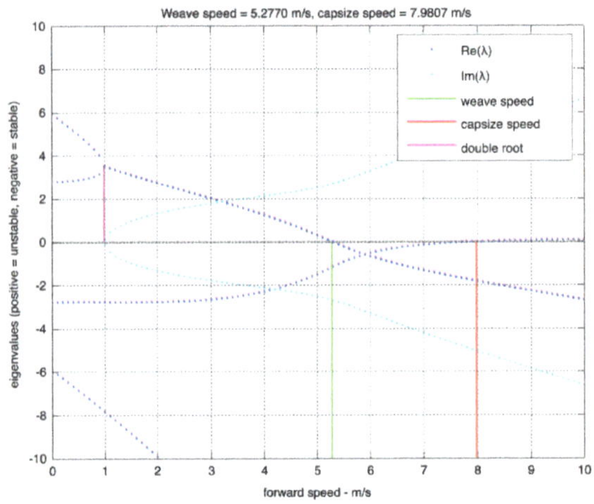

Eigenvalues plotted against forward speed for a typical utility bicycle simplified to have knife-edge wheels that roll without slip.

There are three forward speeds that can be identified in the plot to the right at which the motion of the bike changes qualitatively:

1. The forward speed at which oscillations begin, at about 1 m/s (2.2 mph) in this example, sometimes called the *double root* speed due to there being a repeated root to the characteristic polynomial (two of the four eigenvalues have exactly the same value). Below this speed, the bike simply falls over as an inverted pendulum does.

2. The forward speed at which oscillations do not increase, where the weave mode eigenvalues switch from positive to negative in a Hopf bifurcation at about 5.3 m/s (12 mph) in this example, is called the *weave speed*. Below this speed, oscillations increase until the uncontrolled bike falls over. Above this speed, oscillations eventually die out.

3. The forward speed at which non-oscillatory leaning increases, where the capsize mode eigenvalues switch from negative to positive in a pitchfork bifurca-

tion at about 8 m/s (18 mph) in this example, is called the *capsize speed.* Above this speed, this non-oscillating lean eventually causes the uncontrolled bike to fall over.

Between these last two speeds, if they both exist, is a range of forward speeds at which the particular bike design is self-stable. In the case of the bike whose eigenvalues are shown here, the self-stable range is 5.3–8.0 m/s (12–18 mph). The fourth eigenvalue, which is usually stable (very negative), represents the castoring behavior of the front wheel, as it tends to turn towards the direction in which the bike is traveling. Note that this idealized model does not exhibit the *wobble or shimmy* and *rear wobble* instabilities described above. They are seen in models that incorporate tire interaction with the ground or other degrees of freedom.

Experimentation with real bikes has so far confirmed the weave mode predicted by the eigenvalues. It was found that tire slip and frame flex are *not important for the lateral dynamics of the bicycle in the speed range up to 6* m/s. The idealized bike model used to calculate the eigenvalues shown here does not incorporate any of the torques that real tires can generate, and so tire interaction with the pavement cannot prevent the capsize mode from becoming unstable at high speeds, as Wilson and Cossalter suggest happens in the real world.

Modes

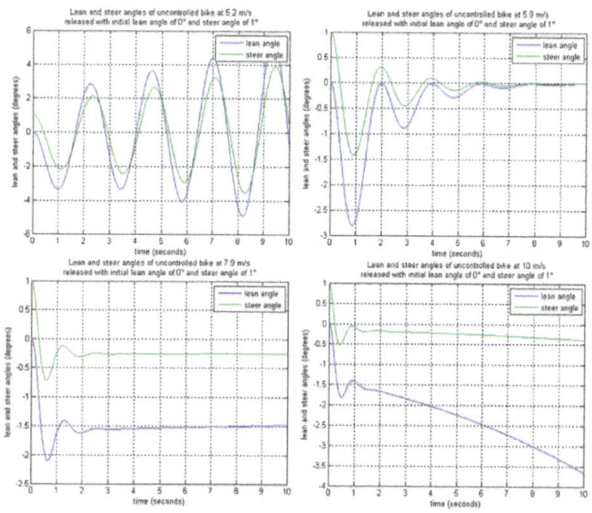

Graphs that show (from left to right, top to bottom) weave instability, self-stability, marginal self-stability, and capsize instability in an idealized linearized model of an uncontrolled utility bicycle.

Bikes, as complex mechanisms, have a variety of modes: fundamental ways that they can move. These modes can be stable or unstable, depending on the bike parameters and its forward speed. In this context, "stable" means that an uncontrolled bike will continue rolling forward without falling over as long as forward speed is maintained. Conversely, "unstable" means that an uncontrolled bike will eventually fall over, even if forward speed is maintained. The modes can be differentiated by the speed at which they switch stability and the relative phases of leaning and steering as the bike experiences that mode. Any bike motion consists of a combination of various amounts of the possible modes, and there are three main modes that a bike can experience: capsize, weave, and wobble. A lesser known mode is rear wobble, and it is usually stable.

Capsize

Capsize is the word used to describe a bike falling over without oscillation. During capsize, an uncontrolled front wheel usually steers in the direction of lean, but never enough to stop the increasing lean, until a very high lean angle is reached, at which point the steering may turn in the opposite direction. A capsize can happen very slowly if the bike is moving forward rapidly. Because the capsize instability is so slow, on the order of seconds, it is easy for the rider to control, and is actually used by the rider to initiate the lean necessary for a turn.

For most bikes, depending on geometry and mass distribution, capsize is stable at low speeds, and becomes less stable as speed increases until it is no longer stable. However, on many bikes, tire interaction with the pavement is sufficient to prevent capsize from becoming unstable at high speeds.

Weave

Weave is the word used to describe a slow (0–4 Hz) oscillation between leaning left and steering right, and vice versa. The entire bike is affected with significant changes in steering angle, lean angle (roll), and heading angle (yaw). The steering is 180° out of phase with the heading and 90° out of phase with the leaning. This AVI movie shows weave.

For most bikes, depending on geometry and mass distribution, weave is unstable at low speeds, and becomes less pronounced as speed increases until it is no longer unstable. While the amplitude may decrease, the frequency actually increases with speed.

Wobble or Shimmy

Wobble, shimmy, tank-slapper, speed wobble, and *death wobble* are all words and phrases used to describe a rapid (4–10 Hz) oscillation of primarily just the front end (front wheel, fork, and handlebars). Also involved is the yawing of the rear frame which may contribute to the wobble when too flexible. This instability occurs mostly at high speed and is similar to that experienced by shopping cart wheels, airplane landing gear, and automobile front wheels. While wobble or shimmy can be easily remedied by adjusting speed, position, or grip on the handlebar, it can be fatal if left uncontrolled.

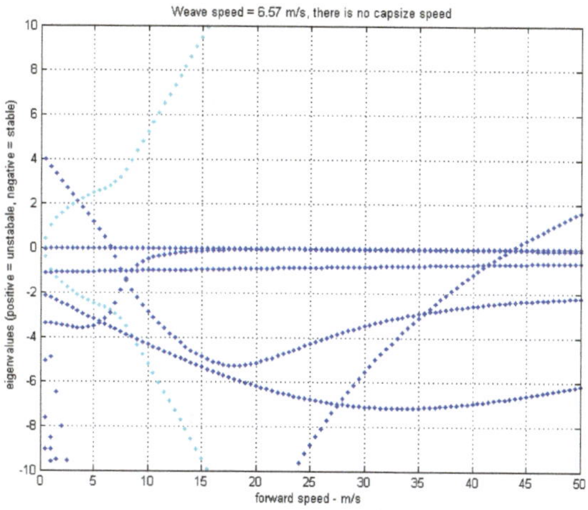

Eigenvalues plotted against forward speed for a motorcycle modeled with frame flexibility and realistic tire dynamics. Additional modes can be seen, such as wobble, which becomes unstable at 43.7 m/s.

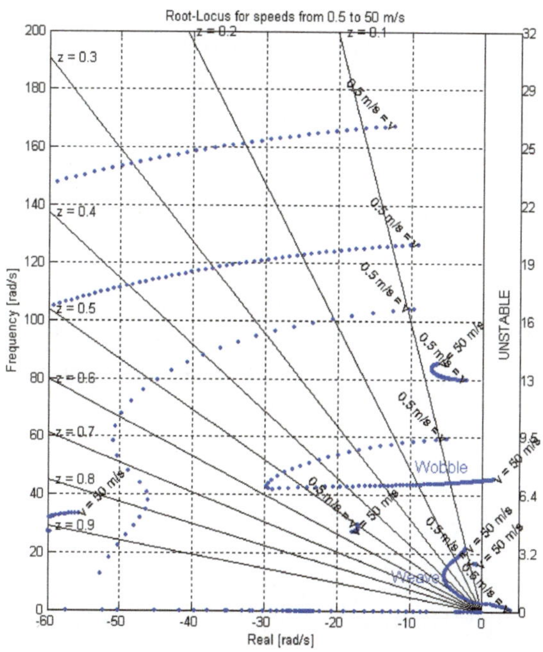

The same eigenvalues as in the figure above, but plotted on a root locus plot. Several additional oscillating modes are visible.

Wobble or shimmy begins when some otherwise minor irregularity, such as fork asymmetry, accelerates the wheel to one side. The restoring force is applied in phase with the progress of the irregularity, and the wheel turns to the other side where the process is repeated. If there is insufficient damping in the steering the oscillation will increase until system failure occurs. The oscillation frequency can be changed by changing the forward speed, making the bike stiffer or lighter, or increasing the stiffness of the steering, of which the rider is a main component.

Rear Wobble

The term *rear wobble* is used to describe a mode of oscillation in which lean angle (roll) and heading angle (yaw) are almost in phase and both 180° out of phase with steer angle. The rate of this oscillation is moderate with a maximum of about 6.5 Hz. Rear wobble is heavily damped and falls off quickly as bike speed increases.

Design Criteria

The effect that the design parameters of a bike have on these modes can be investigated by examining the eigenvalues of the linearized equations of motion. For more details on the equations of motion and eigenvalues.

The lateral and torsional stiffness of the rear frame and the wheel spindle affects wobble-mode damping substantially. Long wheelbase and trail and a flat steering-head angle have been found to increase weave-mode damping. Lateral distortion can be countered by locating the front fork torsional axis as low as possible.

Cornering weave tendencies are amplified by degraded damping of the rear suspension. Cornering, camber stiffnesses and relaxation length of the rear tire make the largest contribution to weave damping. The same parameters of the front tire have a lesser effect. Rear loading also amplifies cornering weave tendencies. Rear load assemblies with appropriate stiffness and damping, however, were successful in damping out weave and wobble oscillations.

One study has shown theoretically that, while a bike leaned in a turn, road undulations can excite the weave mode at high speed or the wobble mode at low speed if either of their frequencies match the vehicle speed and other parameters. Excitation of the wobble mode can be mitigated by an effective steering damper and excitation of the weave mode is worse for light riders than for heavy riders.

Riding on Treadmills and Rollers

Riding on a treadmill is theoretically identical to riding on stationary pavement, and physical testing has confirmed this. Treadmills have been developed specifically for indoor bicycle training. Riding on rollers is still under investigation.

Other Hypotheses

Although bicycles and motorcycles can appear to be simple mechanisms with only four major moving parts (frame, fork, and two wheels), these parts are arranged in a way that makes them complicated to analyze. While it is an observable fact that bikes can be ridden even when the gyroscopic effects of their wheels are canceled out, the hypothesis that the gyroscopic effects of the wheels are what keep a bike upright is common in print and online.

Examples in print:

- "Angular momentum and motorcycle counter-steering: A discussion and demonstration", A. J. Cox, *Am. J. Phys.* 66, 1018–1021 ~1998
- "The motorcycle as a gyroscope", J. Higbie, *Am. J. Phys.* 42, 701–702
- *The Physics of Everyday Phenomena*, W. T. Griffith, McGraw–Hill, New York, 1998, pp. 149–150.
- *The Way Things Work.*, Macaulay, Houghton-Mifflin, New York, NY, 1989

Longitudinal Dynamics

Bikes may experience a variety of longitudinal forces and motions. On most bikes, when the front wheel is turned to one side or the other, the entire rear frame pitches forward slightly, depending on the steering axis angle and the amount of trail. On bikes with suspensions, either front, rear, or both, trim is used to describe the geometric configuration of the bike, especially in response to forces of braking, accelerating, turning, drive train, and aerodynamic drag.

The load borne by the two wheels varies not only with center of mass location, which in turn varies with the amount and location of passengers and luggage, but also with acceleration and deceleration. This phenomenon is known as load transfer or weight transfer, depending on the author, and provides challenges and opportunities to both riders and designers. For example, motorcycle racers can use it to increase the friction available to the front tire when cornering, and attempts to reduce front suspension compression during heavy braking has spawned several motorcycle fork designs.

The net aerodynamic drag forces may be considered to act at a single point, called the center of pressure. At high speeds, this will create a net moment about the rear driving wheel and result in a net transfer of load from the front wheel to the rear wheel. Also, depending on the shape of the bike and the shape of any fairing that might be installed, aerodynamic lift may be present that either increases or further reduces the load on the front wheel.

Stability

Though longitudinally stable when stationary, a bike may become longitudinally unstable under sufficient acceleration or deceleration, and Euler's second law can be used to analyze the ground reaction forces generated. For example, the normal (vertical) ground reaction forces at the wheels for a bike with a wheelbase L and a center of mass at height h and at a distance b in front of the rear wheel hub, and for simplicity, with both wheels locked, can be expressed as:

$N_r = mg\left(\frac{L-b}{L} - \mu\frac{h}{L}\right)$ for the rear wheel and $N_f = mg\left(\frac{b}{L} + \mu\frac{h}{L}\right)$ for the front wheel.

The frictional (horizontal) forces are simply

$F_r = \mu N_r$ for the rear wheel and $F_f = \mu N_f$ for the front wheel,

where μ is the coefficient of friction, m is the total mass of the bike and rider, and g is the acceleration of gravity. Therefore, if

$$\mu \geq \frac{L-b}{h}$$

which occurs if the center of mass is anywhere above or in front of a line extending back from the front wheel contact patch and inclined at the angle

$$\theta = \tan^{-1}\left(\frac{1}{\mu}\right)$$

above the horizontal, then the normal force of the rear wheel will be zero (at which point the equation no longer applies) and the bike will begin to flip or loop forward over the front wheel.

On the other hand, if the center of mass height is behind or below the line, such as on most tandem bicycles or long-wheel-base recumbent bicycles, as well as cars, it is less likely that the front wheel can generate enough braking force to flip the bike. This means they can decelerate up to nearly the limit of adhesion of the tires to the road, which could reach 0.8 g if the coefficient of friction is 0.8, which is 40% more than an upright bicycle under even the best conditions. *Bicycling Science* author David Gordon Wilson points out that this puts upright bicyclists at particular risk of causing a rear-end collision if they tailgate cars.

Similarly, powerful motorcycles can generate enough torque at the rear wheel to lift the front wheel off the ground in a maneuver called a wheelie. A line similar to the one described above to analyze braking performance can be drawn from the rear wheel contact patch to predict if a wheelie is possible given the available friction, the center of mass location, and sufficient power. This can also happen on bicycles, although there is much less power available, if the center of mass is back or up far enough or the rider lurches back when applying power to the pedals.

Of course, the angle of the terrain can influence all of the calculations above. All else remaining equal, the risk of pitching over the front end is reduced when riding up hill and increased when riding down hill. The possibility of performing a wheelie increases when riding up hill, and is a major factor in motorcycle hillclimbing competitions.

Braking According to Ground Conditions

no braking

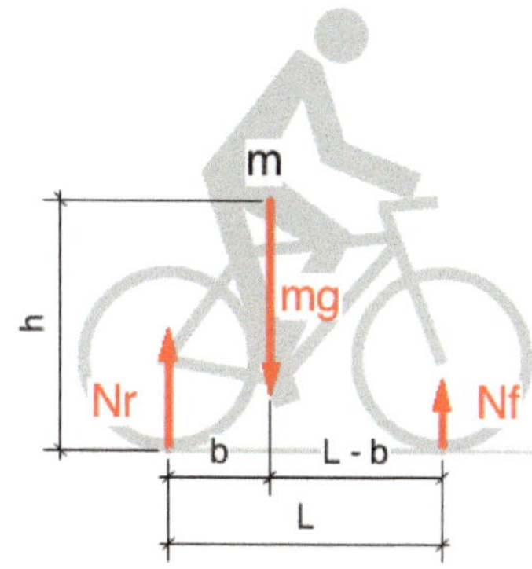

With no braking, on a bicycle *m* is usually approximately over the bottom bracket

When braking, the rider in motion is seeking to change the speed of the combined mass *m* of rider plus bike. This is a negative acceleration *a* in the line of travel. *F=ma*, the acceleration *a* causes an inertial forward force *F* on mass *m*. The braking *a* is from an initial speed *u* to a final speed *v*, over a length of time *t*. The equation $u - v = at$ implies that the greater the acceleration the shorter the time needed to change speed. The stopping distance *s* is also shortest when acceleration *a* is at the highest possible value compatible with road conditions: the equation $s = ut + 1/2\,at^2$ makes *s* low when *a* is high and *t* is low.

How much braking force to apply to each wheel depends both on ground conditions and on the balance of weight on the wheels at each instant in time. The total braking force cannot exceed the gravity force on the rider and bike times the coefficient of friction μ of the tire on the ground. $mg\mu >= Ff + Fr$. A skid occurs if the ratio of either *Ff* over *Nf* or *Fr* over *Nr* is greater than μ, with a rear wheel skid having less of a negative impact on lateral stability.

When braking, the inertial force *ma* in the line of travel, not being co-linear with *f*, tends to rotate *m* about *f*. This tendency to rotate, an overturning moment, is resisted by a moment from *mg*.

light braking

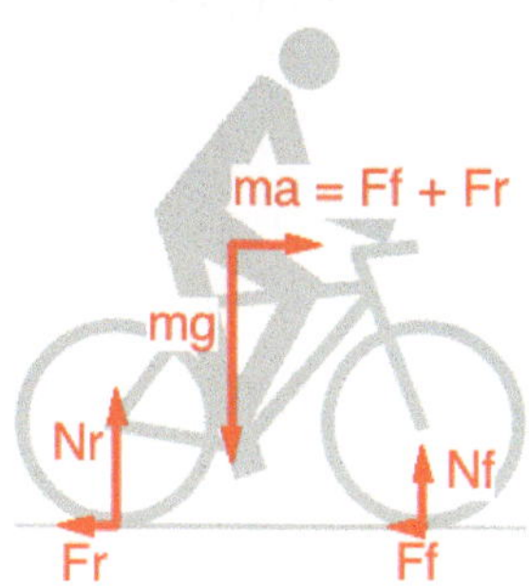

In light braking, *Nr* is still significant so *Fr* can contribute towards braking. *Nr* decreases as *ma* increases

Taking moments about the front wheel contact point at an instance in time:

- When there is no braking, mass *m* is typically above the bottom bracket, about 2/3 of the way back between the front and rear wheels, with *Nr* thus greater than *Nf*.
- In constant light braking, whether because an emergency stop is not required or because poor ground conditions prevent heavy braking, much weight still rests on the rear wheel, meaning that *Nr* is still large and *Fr* can contribute towards *a*.
- As braking *a* increases, *Nr* and *Fr* decrease because the moment *mah* increases with *a*. At maximum constant *a*, clockwise and anti-clockwise moments are equal, at which point *Nr* = 0. Any greater *Ff* initiates a stoppie.

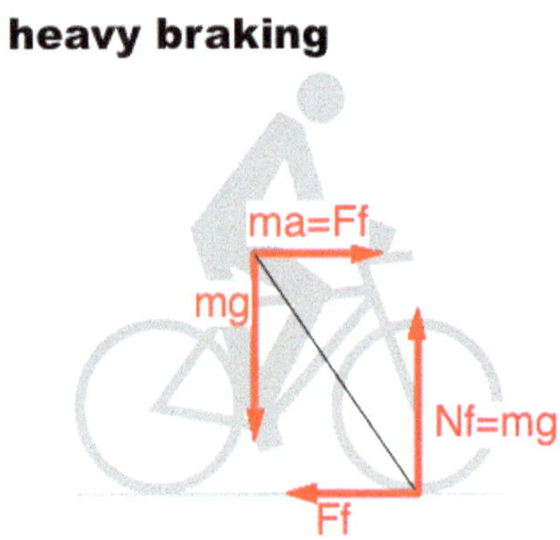

At maximum braking, *Nr* = 0

Other factors:

- Downhill it is much easier to topple over the front wheel because the incline moves the line of *mg* closer to *f*. To try to reduce this tendency the rider can stand back on the pedals to try to keep *m* as far back as possible.
- When braking is increasing the center of mass *m* may move forward relative to the front wheel, as the rider moves forward relative to the bike, and, if the bike has suspension on the front wheel, the front forks compress under load, changing the bike geometry. This all puts extra load on the front wheel.
- At the end of a brake maneuver, as the rider comes to a halt, the suspension decompresses and pushes the rider back.

Values for μ vary greatly depending on a number of factors:

- The material that the ground or road surface is made of.
- Whether the ground is wet or dry.
- The smoothness or roughness of the ground.

- The firmness or looseness of the ground.
- The speed of the vehicle, with friction reducing above 30 mph (50kph).
- Whether friction is rolling or sliding, with sliding friction at least 10% below peak rolling friction.

Braking

A motorcyclist performing a stoppie.

Most of the braking force of standard upright bikes comes from the front wheel. As the analysis above shows, if the brakes themselves are strong enough, the rear wheel is easy to skid, while the front wheel often can generate enough stopping force to flip the rider and bike over the front wheel. This is called a *stoppie* if the rear wheel is lifted but the bike does not flip, or an *endo* (abbreviated form of *end-over-end*) if the bike flips. On long or low bikes, however, such as cruiser motorcycles and recumbent bicycles, the front tire will skid instead, possibly causing a loss of balance. Assuming no loss of balance, it is possible to calculate optimum braking performance depending on the bike's geometry, the location of center of gravity of bike and rider, and the maximum coefficient of friction.

In the case of a front suspension, especially telescoping fork tubes, the increase in downward force on the front wheel during braking may cause the suspension to compress and the front end to lower. This is known as *brake diving*. A riding technique that takes advantage of how braking increases the downward force on the front wheel is known as *trail braking*.

Front Wheel Braking

The limiting factors on the maximum deceleration in front wheel braking are:

- the maximum, limiting value of static friction between the tire and the ground, often between 0.5 and 0.8 for rubber on dry asphalt,

- the kinetic friction between the brake pads and the rim or disk, and
- pitching or looping (of bike and rider) over the front wheel.

For an upright bicycle on dry asphalt with excellent brakes, pitching will probably be the limiting factor. The combined center of mass of a typical upright bicycle and rider will be about 60 cm (24 in) back from the front wheel contact patch and 120 cm (47 in) above, allowing a maximum deceleration of 0.5 *g* (5 m/s² or 16 ft/s²). If the rider modulates the brakes properly, however, pitching can be avoided. If the rider moves his weight back and down, even larger decelerations are possible.

Front brakes on many inexpensive bikes are not strong enough so, on the road, they are the limiting factor. Cheap cantilever brakes, especially with "power modulators", and Raleigh-style side-pull brakes severely restrict the stopping force. In wet conditions they are even less effective. Front wheel slides are more common off-road. Mud, water, and loose stones reduce the friction between the tire and trail, although knobby tires can mitigate this effect by grabbing the surface irregularities. Front wheel slides are also common on corners, whether on road or off. Centripetal acceleration adds to the forces on the tire-ground contact, and when the friction force is exceeded the wheel slides.

Rear-wheel Braking

The rear brake of an upright bicycle can only produce about 0.25 *g* (~2.5 m/s²) deceleration at best, because of the decrease in normal force at the rear wheel as described above. All such bikes with only rear braking are subject to this limitation: for example, bikes with only a coaster brake, and fixed-gear bikes with no other braking mechanism. There are, however, situations that may warrant rear wheel braking.

- Slippery surfaces or bumpy surfaces. Under front wheel braking, the lower coefficient of friction may cause the front wheel to skid which often results in a loss of balance.
- Front flat tire. Braking a wheel with a flat tire can cause the tire to come off the rim which greatly reduces friction and, in the case of a front wheel, result in a loss of balance.
- Front brake failure.
- Recumbent bicycles. Long-wheelbase recumbents require a good rear brake as the CG is near the rear wheel.

Braking Technique

Expert opinion varies from "use both levers equally at first" to "the fastest that you can stop any bike of normal wheelbase is to apply the front brake so hard that the rear wheel is just about to lift off the ground," depending on road conditions, rider skill level, and desired fraction of maximum possible deceleration.

Suspension

Mountain bike rear suspension

Bikes may have only front, only rear, full suspension or no suspension that operate primarily in the central plane of symmetry; though with some consideration given to lateral compliance. The goals of a bike suspension are to reduce vibration experienced by the rider, maintain wheel contact with the ground, and maintain vehicle trim. The primary suspension parameters are stiffness, damping, sprung and unsprung mass, and tire characteristics. Besides irregularities in the terrain, brake, acceleration, and drive-train forces can also activate the suspension as described above. Examples include bob and pedal feedback on bicycles, the shaft effect on motorcycles, and squat and brake dive on both.

Vibration

The study of vibrations in bikes includes its causes, such as engine balance, wheel balance, ground surface, and aerodynamics; its transmission and absorption; and its effects on the bike, the rider, and safety. An important factor in any vibration analysis is a comparison of the natural frequencies of the system with the possible driving frequencies of the vibration sources. A close match means mechanical resonance that can result in large amplitudes. A challenge in vibration damping is to create compliance in certain directions (vertically) without sacrificing frame rigidity needed for power transmission and handling (torsionally). Another issue with vibration for the bike is the possibility of failure due to material fatigue Effects of vibration on riders include discomfort, loss of efficiency, Hand-Arm Vibration Syndrome, a secondary form Raynaud's disease, and whole body vibration. Vibrating instruments may be inaccurate or difficult to read.

In Bicycles

The primary cause of vibrations in a properly functioning bicycle is the surface over which it rolls. In addition to pneumatic tires and traditional bicycle suspensions, a variety of techniques have been developed to damp vibrations before they reach the rider. These include materials, such as carbon fiber, either in the whole frame or just key components such as the front fork, seatpost, or handlebars; tube shapes, such as curved seat stays; and special inserts, such as Zertz by Specialized, and Buzzkills by Bontrager.

In Motorcycles

In addition to the road surface, vibrations in a motorcycle can be caused by the engine and wheels, if unbalanced. Manufacturers employ a variety of technologies to reduce or damp these vibrations, such as engine balance shafts, rubber engine mounts, and tire weights. The problems that vibration causes have also spawned an industry of after-market parts and systems designed to reduce it. Add-ons include handlebar weights, isolated foot pegs, and engine counterweights. At high speeds, motorcycles and their riders may also experience aerodynamic flutter or buffeting. This can be abated by changing the air flow over key parts, such as the windshield.

Experimentation

A variety of experiments have been performed in order to verify or disprove various hypotheses about bike dynamics.

- David Jones built several bikes in a search for an unrideable configuration.
- Richard Klein built several bikes to confirm Jones's findings.
- Richard Klein also built a "Torque Wrench Bike" and a "Rocket Bike" to investigate steering torques and their effects.
- Keith Code built a motorcycle with fixed handlebars to investigate the effects of rider motion and position on steering.
- Schwab and Kooijman have performed measurements with an instrumented bike.
- Hubbard and Moore have performed measurements with an instrumented bike.[

Motorcycle Braking Systems

Motorcycle braking systems have varied throughout time, as motorcycles evolved from bicycles with an engine attached, to the 220 mph (350 km/h) prototype motorcycles seen racing in MotoGP. Most systems work by converting kinetic energy into thermal energy (heat) by friction. On motorcycles, approximately 70% of the braking effort is performed by the front brake. This however can vary for individual motorcycles; longer-wheelbase types having more weight biased rearward, such as cruisers and tourers, can have a `greater effort applied by the rear brake. In contrast, sports bikes with a shorter wheelbase and more vertical fork geometry can tolerate higher front braking loads. For these reasons, motorcycles tend to have a vastly more powerful front brake compared to the rear.

Aprilia Tuono R Front brake system

History

Early motorcycles which were essentially a bicycle with a motor attached and did not have any braking system beyond slowing the motorcycle down and putting a foot out. One of the first motorcycles to have any sort of braking mechanism was made by Steffey Motorcycles of Philadelphia in 1902. This used a spoon brake operating on the front wheel only. Around 1909 band brakes were introduced, these used a band contracting round the outside of a drum.

Spoon front brake operated by handlebar lever and connecting rod on a 1899 Royal Riley Tricycle at the Heritage Motor Centre, Gaydon, England

Douglas motorcycles were available with Research Association disc brakes front and rear on their 1923 RA model, sometimes called TT model, with Freddie Dixon's 1923 sidecar TT-winning machine of that type also having a passenger-operated disc brake for the sidecar wheel.

Drum brakes first started to be fitted to motorcycles in the 1920's and the basic design continues to be used to this day. Originally used for braking both the front and rear wheels, drum brakes have largely been superseded by disc brakes or are used for rear-braking only.

The Lambretta TV125 Series 3 was the first modern two-wheeled production machine to be fitted with a disc brake. MV Augusta followed with their limited production 600 cc touring motorcycle announced in 1965 which had a pair of cable-actuated Campagnolo front disc brakes. In 1969 Honda released the world's first mass produced bike to have a disc brake. The CB750 had a hydraulically operated single piston sliding caliper with a solid front disc.

In 1989 BMW released the first motorcycle to be equipped with anti-lock brakes (ABS). The system fitted to the BMW K100 LT weighed significantly more than the light weight systems fitted to modern motorcycles.

Drum brakes

Drum brakes have a self servo effect. The most common design is a leading-tailing design. More exotic design had four, eight or sixteen shoes. Some motorcycles used finned and/or vented housings for additional cooling, the first of which was AJS.

Torque reaction arm transfer the torque from the brake to fixed point to relieve the suspension reacting to the braking force i.e. diving.

Honda RCB with a front ventilated drum brake from Italian accessories manufacturer Grimeca.

Disc Brakes

Early cast wheels were not easily compatible with drum brakes.

Sintered pads offer improved heat up time and better wet weather performance. Asbestos was used in the brake pads but its use declined when the negative health impact was discovered.

Brake *feel*, rotor size.

Yamaha GTS1000 front swingarm and brake assembly.

The front suspension on the Yamaha GTS1000 released in 1993 was a single sided swingarm that amongst other characteristics aimed to reduce diving under braking. The design only allowed for a single front disc brake so a comparatively large 330mm disc was mated to a six-piston caliper, a world first on a production bike.

Radial master cylinders.

Brake levers, hole in for aerodynamics on race bikes.

Braided brake hoses.

Discs

Floating discs have better disc centering with a fixed caliper. A floating disc also avoids disc warping and reduces heat transfer to the wheel hub. Lambretta were the first manufacturer to use floating discs on a volume production motorcycle.

Wavy discs place the mass closer to the axle for reduced inertia, better heat dissipation and lower weight. They were originally developed for Motorcross bikes however they have since seen use on road going motorcycles as well.

Sliding Versus Fixed

Single action calipers have brake pistons that operate on one side only. This type has a floating pin design which allows the piston side of the caliper to push onto the disc surface and pull the other side into contact as well. Dual action, or opposed calipers, are fixed into position to reduce flex. Dual action have opposed pistons either side of the disc to greatly improve the piston area.

Sliding twin piston caliper on a Honda CB400NC

Multi Piston Calipers

Larger rotors can be used to increase braking force, but this also increases weight and inertia. To overcome this brake manufacturers developed calipers four, six and even eight pistons. Increasing the number of pistons increases the swept piston area, allowing for longer, narrower brake pads and smaller discs.

Caliper Mounting Techniques

Front fork leg mounted brake caliper

Early disc brake calipers were located in front of the fork leg on top of the disc. This position placed the caliper in an area of high air flow for better heat removal. Calipers on most modern motorcycles are mounted to the rear of the fork leg. This has the advantage of reducing the angular momentum of the fork assembly and improving low speed handling. Radial mounted calipers.

Inboard Brakes

In a bid to improve wet weather braking performance, Honda fitted inboard ventilated disc brakes to models such as the VF400F and CBX550F. This saw the front brake assembly enclosed in a vented aluminium hub. The caliper was mounted onto the hub and gripped the disc from the outside. The purpose of inboard discs was to keep the brake assembly dry and allowed the use of cast iron ventilated discs because the shrouding covered any unsightly surface rust on the disc. The system would prove to be short lived with all successive models reverting to the standard uncovered layout thereafter.

Caliper Construction

Yamaha FJR1300 "blue dot" caliper, the same as utilised on many other Yamaha models

Traditionally opposing piston brake calipers are made in two halves which are then bolted together. The advantages of this method are ease and the speed of which the pieces can be made. The disadvantage is that they are less resistant to flex under load. Calipers machined from a single piece of metal are stronger, but much more expensive to make so some manufacturers such as Sumitomo cast the caliper body as one piece and then machine the piston bores externally with the holes created plugged after assembly. This type of caliper design has been used extensively by Yamaha on a wide range of models from its flagship Yamaha YZF-R1 to the more budget Yamaha FZS600 Fazer. Monobloc Brembo first used on the Ducati 1098.

Perimeter Brakes

Buell Motorcycle Company introduced a rim-mounted "zero torsional load" disc brake that was claimed to reduce unsprung weight in the wheel-brake system, including lighter wheel spokes that carried no braking load. This style is generically termed a "perimeter brake" for its point of attachment to the wheel, and had been used in smaller numbers by other manufacturers before Buell. They can rarely be found on custom motorcycle spoked wheels.

Carbon Carbon Disc Brakes

Carbon front brakes on the 2011 Suzuki MotoGP machine

Carbon carbon brakes are used in Moto GP. The friction between the carbon discs and carbon pads produces vast amounts of braking force; far greater braking force than conventional steel disc set-ups. They operate have very high operating temperatures, typically between 400-1000 °C. Continuous late braking can overheat carbon carbon brakes as they oxidize and wear out faster. The high temperatures needed for carbon carbon brakes to be effective means that they cannot be used in the wet. For wet races the riders have to switch a conventional steel brake set-up. Aluminium lithium alloy, banned for 2015 season.

Alternative Systems

Drag-racing motorcycles can reach speeds up to 246 mph (396 km/h) over the course of the ¼ mile (400 m) race and can use disc brakes in conjunction with parachutes to slow them after the timed run. The Streamliner class of motorcycle land speed record machines are mandated to have parachutes fitted if the motorcycle is capable of exceeding 250 miles per hour (400 km/h).

Electric motorcycles can use regenerative braking to both slow the motorcycle down and recharge the batteries at the same.

KTM secretly raced a machine with a KERS style regenerative braking system during the 2008 Valenican 125 cc Grand Prix. Although deemed in contravention of the rules, the use of a KERS system preceded its use in Formula 1.

Suspension (Motorcycle)

A motorcycle's suspension serves a dual purpose: contributing to the vehicle's handling and braking, and providing safety and comfort by keeping the vehicle's passengers comfortably isolated from road noise, bumps and vibrations.

The typical motorcycle has a pair of fork tubes for the front suspension, and a swingarm with one or two shock absorbers for the rear suspension.

Front Suspension

Vincent Black Lightning with Girdraulic front suspension

The most common form of front suspension for a modern motorcycle is the telescopic fork. Other fork designs are girder forks, suspended on sprung parallel links (not common since the 1940s) and bottom leading link designs, not common since the 1960s.

Some manufacturers (e.g. Greeves) used a version of the swinging arm for front suspension on their motocross designs. A single-sided version of the idea is also used in motor scooters such as the Vespa.

The Hub-center steering as developed by Ascanio Rodorigo, on a concept associated to Massimo Tamburini is a complex front swingarm alternative system that entails suspension and steering, as seen in projects such as Bimota Tesi and Vyrus motorcycles.

Telescopic Forks

Scott produced a motorcycle with telescopic forks in 1908, and would continue to use them until 1931. In 1935 BMW became the first manufacturer to produce a motorcycle with hydraulically damped telescopic forks. Most motorcycles today use telescopic forks for the front suspension. The forks can be most easily understood as simply large hydraulic shock absorbers with internal coil springs. They allow the front wheel to react to imperfections in the road while isolating the rest of the motorcycle from that motion.

Telescopic forks on a 1969 BMW

The top of the forks are connected to the motorcycle's frame in a triple tree clamp (known as a "yoke" in the UK), which allows the forks to be turned in order to steer the motorcycle. The bottom of the forks is connected to the front wheel's axle.

On conventional telescopic forks, the lower portion or *fork bodies* (*"fork sliders"* in the UK), slide up and down the *fork tubes* (*"fork stanchions"* in the UK). The fork tubes must be mirror-smooth to seal the fork oil inside the fork. Some fork tubes, especially on early roadsters and off-road motorcycles, are enclosed in plastic protective "gaiters.

"Upside-down" (USD) forks, also known as inverted forks, are installed inverted compared to conventional telescopic forks. The slider bodies are at the top, fixed in the triple clamps, and the stanchion tubes are at the bottom, fixed to the axle. This USD arrangement has two advantages: (i) it decreases the unsprung weight of the motorcycle; and (ii) it increases torsional stiffness, which can improve handling. Two disadvantages of USD forks are: (i) they are more expensive than conventional telescopic forks; and (ii) they are liable to lose all their damping oil should an oil seal fail. USD forks are typically found on sportbikes, though the Honda Valkyrie featured USD forks.

Pre-load Adjustment

Motorcycle suspensions are designed so that the springs are always under compression, even when fully extended. Pre-load is used to adjust the initial position of the suspension with the weight of the motorcycle and rider acting on it.

The difference between the fully extended length of the suspension and the length compressed by the weight of the motorcycle and rider is called "total sag" or "race sag". Total sag is set to optimize the initial position of the suspension to avoid bottoming out or topping out under normal riding conditions. "Bottoming out" occurs when the suspension is compressed to the point where it mechanically cannot compress any more. Topping out occurs when the suspension extends fully and cannot mechanically extend any more. Increasing pre-load increases the initial force on the spring thereby reducing total sag. Decreasing pre-load decreases the initial force in the spring thereby increasing total sag.

Some motorcycles allow adjustment of pre-load by changing the air pressure inside the forks. Valves at the top of the forks allow air to be added or released from the fork. More air pressure gives more pre-load, and vice versa.

Fork Damping

Basic fork designs use a simple *damper-rod* system, in which damping is controlled by the passage of fork oil through an orifice. Though cheap to manufacture, it is hard to tune such forks, as they tend to give too little damping at low slider speeds, yet too much damping at higher slider speeds. Any adjustment setting will always be a compromise, giving both over-mushy and over-stiff damping. Since forks act as hydraulic dampers, changing the weight of the fork oil will alter the damping rate. Some telescopic forks have external adjustments for damping.

A more sophisticated approach is the *cartridge fork*, which use internal cartridges with a valving system. Damping at low slider speeds is controlled by a much smaller orifice, but damping at higher slider speeds is controlled by a system of flexible shims, which act as a bypass valve for the fork oil. This valve has a number of such shims of varying thicknesses that cover the orifices in the valve to control the damping of the fork on high and medium speed bumps.

Some of the shims (or "leaf springs") lift with little force allowing fluid to flow through the orifice. Other springs require greater force to lift and allow flow. This gives the fork *digressive* damping, allowing it to be stiff over small bumps, yet relatively softer over larger bumps. Also, the springs (or shims) only allow flow in one direction, so one set of springs controls compression damping, and another rebound damping. This allows the dampings to be set separately.

Cartridge emulators are aftermarket parts that make damper-rod forks behave virtually as cartridge forks. The damping orifice in the damper-rod is made so large that it has virtually no effect on damping, and instead an "emulator" takes over the damping function. The emulator has a very small orifice for low fork-speed damping, and an adjustable shim-stack for high fork-speed damping.

Gas-charged cartridge forks, which became available in 2007, consist of gas-charged cartridges fitted within standard forks. This kit is suitable for supersport classes of racing, where regulations prohibit a complete fork replacement but allow modification of the original forks.

Brake Dive

Applying the brakes of a moving motorcycle increases the load borne by the front wheel and decrease the load borne by the rear wheel due to a phenomenon called load transfer.

BMW's 1955-1969 Earles fork eliminated and reversed brake dive

If the motorcycle is equipped with telescopic forks, the added load on the front wheel is transmitted through the forks, which compress. This shortening of the forks causes the front end of the bike to move lower, and this is called brake dive. telescopic forks are particularly prone to this, unlike leading link designs.

Brake dive can be disconcerting to the rider, who may feel like he or she is about to be thrown over the front of the motorcycle. If the bike dives so far as to bottom out the front forks, it can also cause handling and braking problems. One of the purposes of a suspension is to help maintain contact between the tire and road. If the suspension has bottomed out, it is no longer moving as it should, and is no longer helping to maintain contact.

While excessive brake dive is disconcerting, and bottoming out can cause loss of traction, a certain amount of brake dive reduces the rake and trail of the motorcycle, allowing it to more easily turn. This is especially important to racers trail braking on entrance to corners.

Brake dive with telescopic forks can be reduced by either increasing the spring rate of the fork springs, or increasing the compression damping of the forks. However, all of these changes make the motorcycle less pleasant to ride on rough roads, since the front end will feel stiffer, in the 1980s various manufacturers attempted to get round this by methods of anti-dive such as:

- ACT: Developed by Marzocchi and fitted to Buell motorcycles such as the Buell RR 1200 (1988).
- ANDF (Anti Nose Dive Forks): This was fitted to a number of Suzuki GSX models and the RG250.
- AVDS (Automatic Variable Damping System): This was fitted to a number of Kawasaki motorcycles.
- NEAS (New Electrically Activated Suspension): As fitted to the Suzuki GSX-R 1100 and GSX-R 750 Limited Edition.

- PDF (Posi Damp Fork): This was fitted to the Suzuki RG500, the GSX1100E/GS1150E and GSX-R 750 and was an enhancement of the previous Anti Dive units (which works by brake fluid pressure closing a valve in the mechanism when the brakes are applied, restricting the flow of damping oil and slowing fork compression). The PDF units works in a similar way, except that they rely entirely on the compression stroke itself. The valves are spring-loaded so if the wheel hits a bump, they bounce off their seats and restore the flow of oil for a moment to allow the suspension to absorb the shock.
- TCS (Travel Control System): Anti-dive system with variable damping. TCS was introduced on the FZ 400 R (1984, only for the Japanese market).
- TRAC (Torque Reactive Anti-dive Control): This was fitted to a number of Honda motorcycles such as the CB1100F, CB1000C, and VFR750F and worked by utilizing a pivoting caliper that activated a valve in the fork leg.

With the advent of cartridge forks, which allow more low speed damping and less high speed damping than the previous damping rod forks, separate anti-dive mechanisms have generally fallen out of use.

Another method to reduce or eliminate brake dive in telescopic forks is to use a reactive link or torque arm to connect the braking components to the motorcycle frame via the triple clamp.

Some fork designs mitigate dive, eliminate it, or even reverse it without affecting the front suspension adversely. The Earles fork is among the latter; when braking the front brake hard, the front end of the motorcycle actually rises. BMW's Telelever fork is designed to nearly eliminate dive, and could have been designed to eliminate it completely if the manufacturer chose to do so. Leading link front forks, such as used on some Ural motorcycles, can also be designed either to reduce or eliminate dive.

Saxon-Motodd (Telelever) Fork

BMW R1200c with Telelever

The Saxon-Motodd (marketed as Telelever by BMW) has an additional swingarm that mounts to the frame and supports the spring. This causes the rake and trail to increase during braking instead of decreasing as with traditional telescopic forks.

Hossack/Fior (Duolever) Fork

The Hossack/Fior (marketed as Duolever by BMW) separates completely the suspension from steering forces. It was developed by Norman Hossack though used by Claude Fior and John Britten on racebikes. Hossack himself described the system as a 'steered upright'. In 2004 BMW announced the K1200S with a new front suspension that is based upon this design.

Single-sided

A single-sided front swingarm suspension was used on the Yamaha GTS1000, introduced in 1993. The GTS used the RADD, Inc. front suspension designed by James Parker. A single sided girder fork was use on the German Imme R100 motorcycle between 1949 and 1951, and the Vespa scooter has a single-sided trailing-link fork. More recently, between 1998 and 2003, the ItalJet "Dragster" scooter also used a single-sided swingarm suspension, though unlike the GTS1000 there was no upper control arm; the upper part of the suspension on the Dragster served only to transmit steering input.

Hub-center Steering

Hub-center steering is characterized by a swingarm that extends from the bottom of the engine/frame to the centre of the front wheel instead of a fork.

The advantages of using a hub-center steering system instead of a more conventional motorcycle fork are that hub-center steering separates the steering, braking, and suspension functions.

With a fork the braking forces are put through the suspension, a situation that leads to the suspension being compressed, using up a large amount of suspension travel which makes dealing with bumps and other road irregularities extremely difficult. As the forks dive the steering geometry of the bike also changes making the bike more nervous, and inversely on acceleration becomes more lazy. Also, having the steering working through the forks causes problems with stiction, decreasing the effectiveness of the suspension. The length of the typical motorcycle fork means that they act as large levers about the headstock requiring the forks, the headstock, and the frame to be very robust adding to the bike's weight.

The endurance racer "Nessie", built by the Mead & Tomkinson racing team, used an adapted version of Difazio hub-center steering, whereby braking forces were directed to the frame via a pivoted fork (rather than through the steering head). This allowed neutral steering and an absence of brake dive.

Rear Suspension

Plunger rear suspension on a BMW R51/3

Early Rear Suspensions

While front suspensions were almost universally adopted before World War I, several manufacturers did not use rear suspension on their bikes until after World War II. However, motorcycles with rear suspension were offered to the public before World War I. Notable among these are the 1913 Indian Single with a swingarm suspended from a leaf spring and the 1913 Pope with wheels supported on a pair of plungers which were each suspended by a coil spring.

Plunger Suspension

Several motorcycles before and immediately after World War II used plunger suspension in which the vertical movement of the rear axle was controlled by plungers suspended by springs.

Notable manufacturers of bikes with plunger suspension include Adler, Ariel, BMW, BSA, Indian, MZ, Saroléa, Norton, Cossack/Ural and Zündapp.

Although plunger suspension could be sophisticated, with springing and damping in both compression and rebound, it had three disadvantages (compared to the almost universal swingarm that succeeded it), as follows: (i) wheel travel was limited, (ii) the wheel could move out of the required vertical axis, and (iii) it was more expensive to produce and maintain.

Swingarms

The basic motorcycle swingarm is a quadrilateral, with one short side connected to the motorcycle's frame with bearings so that it can pivot. The other short side is the rear axle around which the rear wheel turns. The long sides are connected to the motorcycle's frame or rear sub-frame with one or two shocks with coil-over springs.

In production motorcycles, swingarms are not exactly rectangular, but their function can be more easily understood by thinking of them as such.

Moto Guzzi's *CA.R.C.*

When a swingarm is present on only one side of the motorcycle, this is known as a *single-sided* swingarm. Notable examples include the Honda VFR800 and the BMW R- and K-series. Single-sided swingarms make rear-wheel removal easier, though they generally increase the unsprung weight of the rear suspension. This is due to the additional material required to give identical torsional rigidity to a conventional (two-sided) swingarm setup. For this reason sports bikes are rarely seen using the setup. Notable exclusions are the Ducati 916 which was intended to be taken endurance racing, the MV Agusta f4 which has a hollow interior for reducing weight (a magnesium version is also available), and the Ducati 1098, which was given a single sided swingarm purely for styling reasons.

BMW's Paralever rear suspension on a R1200GS

On many shaft-drive motorcycles the drive shaft is contained in one of the long sides of the swingarm. Notable examples include all post-1955 BMW models prior to BMW's use of the single-sided swingarms, Urals, many Moto Guzzi twins, the Honda Goldwing, the Yamaha XS Eleven, and the Yamaha FJR1300.

The BMW R- and K-series combine a shaft-drive contained in the swing arm with a single-sided swingarm, and the combination is marketed as the *Paralever*. Newer Moto

Guzzi motorcycles use a similar arrangement marketed as the *CA.R.C.* ("CArdano Reattivo Compatto" - Compact Reactive Shaft Drive).

For motorcycles with chain drives, the rear axle can usually be adjusted forward and back in relation to the swingarm, to adjust chain tension, but a few models (such as 1971/72 Triumph and BSA four-strokes) make the adjustment at the swingarm's pivot point.

Shock Absorbers

The hydraulic shock absorbers used on the rear suspensions of motorcycles are essentially the same as those used in other vehicle applications.

Motorcycle shocks do differ slightly in that they nearly always use a coil-over spring. In other words, the spring for the rear suspension is a coil spring that is installed over, or around, the shock.

In terms of adjustment, rear shocks span the range from no adjustments whatsoever to pre-load adjustments only to racing shocks with adjustments for length, pre-load, and four different kinds of damping. Most shocks have internal oil reservoirs, but some have external ones, and some offer air-assisted damping.

A number of companies offer custom-built rear shocks for motorcycles. These shocks are assembled for a specific motorcycle and rider combination, taking into account the characteristics of the motorcycle, the weight of the rider, and the rider's preferred riding style/aggressiveness.

Twin Shock Absorbers

Twinshock refers to motorcycles that have two shock absorbers. Generally, this term is used to denote a particular era of motorcycles, and is most frequently used when describing off-road motorcycles.

During the late 1970s and 1980s, motorcycle rear suspension design and performance underwent tremendous advances. The primary goal and result of these advances were increased rear wheel travel, as measured in the how far the rear wheel could move up and down. Before this period of intense focus on rear suspension performance, most off-road motorcycles had rear wheel travel of about 3.5–4 inch (9–10 cm). At the end of this period, most of these motorcycles had rear wheel travel of approximately 12 inch (30 cm). At the beginning of this period, various rear suspension designs were used to reach this degree of performance. However, by the end of this period, a design consisting of using only one shock absorber (instead of two) was universally accepted and used. Motorcycles with only one shock absorber are called monoshock motorcycles. The performance of monoshock motorcycles was vastly superior to twin shock motorcycles. Accordingly, this design distinction is readily used to categorize motorcycles. With the exception of Bentley and Draper system (New Imperial and Brough machines) and the HRD (later Vincent) system both developed and patented in the 1920s it is only since the 1980s that mono-shock motorcycles have been the norm, the term "twinshock" is now used to categorize vintage motorcycles. This distinction is important in that it provides classes used for vintage motorcycle competition. For example, vintage motocross races are held for older motocross motorcycles. To prevent the better-performing monoshock motorcycles from dominating the competition, there are separate competition classes for monoshock and twinshock motorcycles, which prevents them from competing directly against each other.

Mono-shock

On a motorcycle with a mono-shock rear suspension, a single shock absorber connects the rear swingarm to the motorcycle's frame. Typically this lone shock absorber is in front of the rear wheel, and uses a linkage to connect to the swing arm. Such linkages are frequently designed to give a rising rate of damping for the rear. Honda refers to its mono-shock designs as *Pro-link* suspensions, Kawasaki as *Uni-Trak*, Suzuki as *Full-Floater*, and Yamaha as *Monocross*. Honda's Unit Pro-Link, used first on the Honda RC211V MotoGP racer, and then on the 2003 Honda CBR600RR sport bike, is intended to isolate the frame and the steering head from undesirable forces transmitted by the rear suspension by having the dampers upper mount contained within the rear swingarm subframe, rather than connecting it to the frame itself.

Motorcycle Fork

1968 BMW R60US with conventional telescopic fork

A motorcycle fork connects a motorcycle's front wheel and axle to its frame, typically via a pair of triple clamps. It typically incorporates the front suspension and front brake, and allows the front wheel to rotate about the steering axis so that the bike may be steered. Handlebars attach to the top clamp.

Unusual "trailing bottom link" on a Honda Rune

The fork and its attachment points on the frame establish the critical geometric parameters of rake and trail, which play a major role in defining how a motorcycle handles and dives during braking.

Variations

A variety of front fork arrangements have been tried during more than one hundred years of motorcycle development, several of which remain available today.

Telescopic

A telescopic fork uses fork tubes which contain the suspension components (coil springs and damper) internally. This is the most common form of fork commercially available. It may or may not include gaiters for protection against abrasive elements on the suspension cylinders. The main advantages of the telescopic fork are that (i) it is simple in design and relatively cheap to manufacture and assemble; (ii) it is lighter than older designs using external components and linkage systems; and (iii) it has a clean and simple appearance that bikers find attractive.

BMW's version of oil-damped telescopic fork, on a 1939 R12

Conventionally, the fork stanchions are at the top, clamped to a triple tree (also called a triple clamp or a yoke), and the sliders are at the bottom, attached to the front wheel spindle. On some modern sport bikes and most off-road bikes, this system is inverted, with "sliders" (complete with the spring/damper unit) at the top, clamped to the yoke, while the stanchions are at the bottom. This is done (i) to reduce unsprung weight by having the heavier components suspended, and (ii) to improve the strength and rigidity of the assembly by having the strong large-diameter "sliders" clamped in the yokes. The inverted system is referred to as an *upside-down fork*, or "USD" for short. A disadvantage of this USD design is that the entire reservoir of damping oil is above the slider seal so that, if the slider seal were to leak, the oil could drain out, rendering any damping ineffective.

Trailing Link

Trailing link fork on a 1928 Indian Big Chief

A trailing link fork suspends the wheel on a link (or links) with a pivot point forward of the wheel axle. Most famously used by Indian Motocycle; it was also used by BMW for its early bikes.

Leading Link

Ural's variant of the leading link fork

A leading link fork suspends the wheel on a link (or links) with a pivot point aft of the wheel axle. Russian Ural motorcycles still use leading link forks on sidecar equipped motorcycles, and aftermarket leading link forks are often installed today on motorcycles when they are outfitted with sidecars. They are also very popular with trikes, improving the handling while steering or braking. The most common example of a leading link fork is that found on the Honda Super Cub.

Springer

Early Harley-Davidson with springer fork

The springer fork is an early type of leading link fork. A springer fork does not have the suspension built into the fork tubes, but instead has it mounted externally, where it may be integrated into the triple tree. This style of fork may be found on antique motorcycles or choppers, and is available today on Harley-Davidson's *Softail Springer*.

While it may have an exposed spring near the triple clamp, a springer fork is distinguishable from a girder fork by its two parallel sets of legs. The rear is firmly fixed to the bottom triple clamp (usually brazed or welded). A short leading link holds the wheel and the forward leg which actuates the springs (usually mounted on the triple clamp).

Earles

Earles fork on a 1952 Douglas Dragonfly

The Earles fork is a variety of leading link fork where the pivot point is behind the front wheel, which is the basis of the Earles' patent. Patented by Englishman Ernest Earles in 1953, the design is constructed of light tubing, with conventional 'shock absorbers' mounted near the front axle. The Earles fork has a very small wheelbase change under braking or under compression, unlike telescopic forks. Their construction is much stronger than teleforks, especially against lateral deflection caused by hard cornering (as when racing), or when cornering with a sidecar. This triangulated fork causes the front end of a motorcycle to rise slightly when braking hard, as the mechanical braking forces rotate 'downward' relative to the fork's pivot point — this action can be disconcerting to riders used to telescopic forks, which have the opposite reaction to braking forces ('brake dive'). Several motorcycle manufacturers licensed the Earles patent forks for racing motorcycles in 1953, such as MV Agusta and BMW Motorcycle, while other companies (such as Douglas motorcycle) used the Earles design on their roadsters or off-road machines. BMW used Earles forks on all their motorcycles between 1955 and 1968.

Girder

One of the earliest types of motorcycle front suspension, the girder fork consists of a pair of uprights attached to the triple clamp by linkages with a spring usually between the top and bottom triple clamps. The design reached a peak in the "Girdraulics" used on the Vincent motorcycle. Girdraulic forks featured forged alloy blades for extra strength and hydraulic damping.

1934 Cotton with girder fork

While both may have an exposed spring near the triple clamp, a girder fork is distinguishable from a springer fork by the wheel being fixed firmly to the (usually a long diamond shape) upright. The pivot points are short links mounted to the top and bottom triple clamps. The spring is usually mounted to the girder and compressed against the upper triple clamp.

Although girder forks are associated with early motorcycles, the British Ariel Ace, a modern sports motorcycle, has an alloy space frame and a matching girder fork. This

Ariel girder fork is just as prone to brake dive as any telescopic fork, but is claimed to be more resistant to torsional twisting.

Saxon-Motodd (Telelever)

Telelever fork on a BMW R1150R

The Saxon-Motodd (marketed as Telelever by BMW) has a wishbone that mounts to the frame and supports the monoshock unit. This relieves the forks of any braking and suspension forces. With a Telelever, the trail and castor angle (rake) increases during braking instead of decreasing as with traditional telescopic forks. BMW's boxer twins have been equipped with Telelever forks since 1994, but some newer boxers like the BMW R nineT have reverted to conventional telescopic forks for aesthetic and packaging reasons.

Hossack/Fior (Duolever)

Duolever front fork

The Hossack/Fior (marketed as Duolever by BMW) completely separates the suspension from steering forces. Developed by Norman Hossack and used by Claude Fior and John Britten on racebikes, Hossack described the system as a 'steered upright'. In 2004 BMW announced the K1200S with a new front suspension that appears to be based upon the design. As of 2006, the Duolever is on the K1200S, K1200R, and K1200GT.

Coaxial Steering Front Suspension

Developed by MotoCzysz for their C1 and awarded United States Patent 7111700 on September 26, 2006. Defined as a motorcycle or bicycle front end having coaxial steering and suspension components, and having telescopic forks. Swing weight of the forks is dramatically reduced by removing their suspension components to the central location, coaxially within the steering tube. Ride height can be adjusted without loosening the forks in the triple clamps.

This particular fork, as implemented on the MotoCzysz C1, also has adjustable trail, from 89 mm to 101 mm.

Non-forks

Yamaha GTS1000 suspension

There have been several attempts to implement front steering and suspension without using anything that could be described as a "fork". Examples include hub-center steering, used as early as 1920 on the Ner-a-Car, and implemented in the 1990s on the Bimota Tesi and the Yamaha GTS1000.

A single-sided girder "fork" was used by the German firm Imme between 1949 and 1951.

Mead & Tomkinson Racing competed in endurance racing in the 1970s with "Nessie," a Laverda-based racing motorcycle with hub-centre steering.

Shock Absorber

Miniature oil filled Coilover shock components for scale cars.

A shock absorber (in reality, a shock "damper") is a mechanical or hydraulic device designed to absorb and damp shock impulses. It does this by converting the kinetic energy of the shock into another form of energy (typically heat) which is then dissipated. Most shock absorbers are a form of dashpot.

Description

Pneumatic and hydraulic shock absorbers are used in conjunction with cushions and springs. An automobile shock absorber contains spring-loaded check valves and orifices to control the flow of oil through an internal piston.

One design consideration, when designing or choosing a shock absorber, is where that energy will go. In most shock absorbers, energy is converted to heat inside the viscous fluid. In hydraulic cylinders, the hydraulic fluid heats up, while in air cylinders, the hot air is usually exhausted to the atmosphere. In other types of shock absorbers, such as electromagnetic types, the dissipated energy can be stored and used later. In general terms, shock absorbers help cushion vehicles on uneven roads.

Vehicle Suspension

In a vehicle, shock absorbers reduce the effect of traveling over rough ground, leading to improved ride quality and vehicle handling. While shock absorbers serve the purpose of limiting excessive suspension movement, their intended sole purpose is to damp spring oscillations. Shock absorbers use valving of oil and gasses to absorb excess energy from the springs. Spring rates are chosen by the manufacturer based on the weight of the vehicle, loaded and unloaded. Some people use shocks to modify spring rates but this is not the correct use. Along with hysteresis in the tire itself, they damp the energy stored in the motion of the unsprung weight up and down. Effective wheel bounce damping may require tuning shocks to an optimal resistance.

Spring-based shock absorbers commonly use coil springs or leaf springs, though torsion bars are used in torsional shocks as well. Ideal springs alone, however, are not shock absorbers, as springs only store and do not dissipate or absorb energy. Vehicles typically employ both hydraulic shock absorbers and springs or torsion bars. In this combination, "shock absorber" refers specifically to the hydraulic piston that absorbs and dissipates vibration. Now, composite suspension system are used mainly in 2 wheelers and also leaf spring are made up of composite material in 4 wheelers.

Early History

In common with carriages and railway locomotives, most early motor vehicles used leaf springs. One of the features of these springs was that the friction between the leaves offered a degree of damping, and in a 1912 review of vehicle suspension the lack of this characteristic in helical springs was the reason it was "impossible" to use them as main springs. However the amount of damping provided by leaf spring friction was limited and variable according to the conditions of the springs, and whether wet or dry. It also operated in both directions. Motorcycle front suspension adopted coil sprung Druid forks from about 1906, and similar designs later added rotary friction dampers, which damped both ways - but they were adjustable (e.g. 1924 Webb forks). These friction disk shock absorbers were also fitted to many cars.

One of the problems with motor cars was the large variation in sprung weight between lightly loaded and fully loaded, especially for the rear springs. When heavily loaded the springs could bottom out, and apart from fitting rubber 'bump stops', there were attempts to use heavy main springs with auxiliary springs to smooth the ride when lightly loaded, which were often called 'shock absorbers'. Realising that the spring and vehicle combination bounced with a characteristic frequency, these auxiliary springs were designed with a different period, but were not a solution to the problem that the spring rebound after striking a bump could throw you out of your seat. What was called for was damping that operated on the rebound.

Although C.L. Horock came up with a design in 1901 that had hydraulic damping, it worked in one direction only. It does not seem to have gone into production right away, whereas mechanical dampers such as the Gabriel Snubber started being fitted in the late 1900s (also the similar Stromberg Anti-Shox). These used a belt coiled inside a device such that it freely wound in under the action of a coiled spring, but met friction when drawn out. Gabriel Snubbers were fitted to an 11.9HP Arrol-Johnston car which broke the 6 hour Class B record at Brooklands in late 1912, and the Automotor journal noted that this snubber might have a great future for racing due to its light weight and easy fitment.

One of the earliest hydraulic dampers to go into production was the Telesco Shock Absorber, exhibited at the 1912 Olympia Motor Show and marketed by Polyrhoe Carburettors Ltd. This contained a spring inside the telescopic unit like the pure spring type 'shock absorbers' mentioned above, but also oil and an internal valve so that the oil damped in the rebound direction. The Telesco unit was fitted at the rear end of the leaf spring, in place of the rear spring to chassis mount, so that it formed part of the springing system, albeit a hydraulically damped part. This layout was presumably selected as it was easy to apply to existing vehicles, but it meant the hydraulic damping was not applied to the action of the main leaf spring, but only to the action of the auxiliary spring in the unit itself.

The first production hydraulic dampers to act on the main leaf spring movement were probably those based on an original concept by Maurice Houdaille patented in 1908 and 1909. These used a lever arm which moved hydraulically damped vanes inside the unit. The main advantage over the friction disk dampers was that it would resist sudden movement but allow slow movement, whereas the rotary friction dampers tended to stick and then offer the same resistance regardless of speed of movement. There appears to have been little progress on commercialising the lever arm shock aborbers until after World War I, after which they came into widespread use, for example as standard equipment on the 1927 Ford Model A.

Types of Vehicle Shock Absorbers

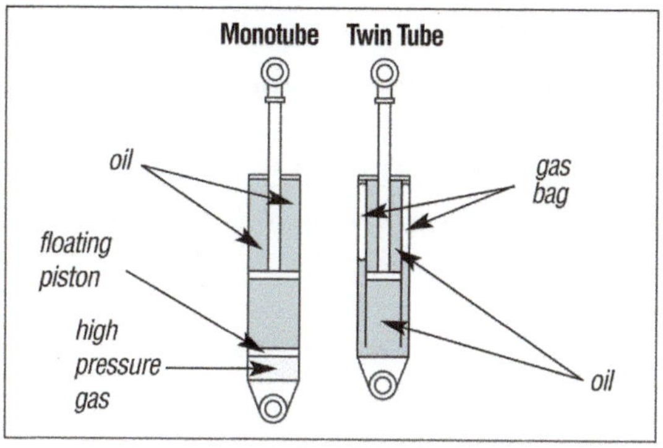

Diagram of the main components of a twin-tube and mono-tube shock absorber

Most vehicular shock absorbers are either twin-tube or mono-tube types with some variations on these themes.

Twin-tube

Basic Twin-tube

Also known as a "two-tube" shock absorber, this device consists of two nested cylindrical tubes, an inner tube that is called the "working tube" or the "pressure tube", and an outer tube called the "reserve tube". At the bottom of the device on the inside is a compression valve or base valve. When the piston is forced up or down by bumps in the road, hydraulic fluid moves between different chambers via small holes or "orifices" in the piston and via the valve, converting the "shock" energy into heat which must then be dissipated.

Twin-tube Gas Charged

Variously known as a "gas cell two-tube" or similarly-named design, this variation represented a significant advancement over the basic twin-tube form. Its overall structure is very similar to the twin-tube, but a low-pressure charge of nitrogen gas is added to the reserve tube. The result of this alteration is a dramatic reduction in "foaming" or "aeration", the undesirable outcome of a twin-tube overheating and failing which presents as foaming hydraulic fluid dripping out of the assembly. Twin-tube gas charged shock absorbers represent the vast majority of original modern vehicle suspensions installations.

Position Sensitive Damping

Often abbreviated simply as "PSD", this design is another evolution of the twin-tube shock. In a PSD shock absorber, which still consists of two nested tubes and still contains nitrogen gas, a set of grooves has been added to the pressure tube. These grooves allow the piston to move relatively freely in the middle range of travel (i.e., the most common street or highway use, called by engineers the "comfort zone") and to move with significantly less freedom in response to shifts to more irregular surfaces when upward and downward movement of the piston starts to occur with greater intensity (i.e., on bumpy sections of roads— the stiffening gives the driver greater control of movement over the vehicle so its range on either side of the comfort zone is called the "control zone"). This advance allowed car designers to make a shock absorber tailored to specific makes and models of vehicles and to take into account a given vehicle's size and weight, its maneuverability, its horsepower, etc. in creating a correspondingly effective shock.

Acceleration Sensitive Damping

The next phase in shock absorber evolution was the development of a shock absorber that could sense and respond to not just situational changes from "bumpy" to "smooth" but to individual bumps in the road in a near instantaneous reaction. This was achieved through a change in the design of the compression valve, and has been termed "acceleration sensitive damping" or "ASD". Not only does this result in a complete disappearance of the "comfort vs. control" tradeoff, it also reduced pitch during vehicle braking and roll during turns. However, ASD shocks are usually only available as aftermarket changes to a vehicle and are only available from a limited number of manufacturers.

Coilover

Coilover shock absorbers are usually a kind of twin-tube gas charged shock absorber around which has been mounted a large metal coil. Though common on motorcycle and scooter rear suspensions, coilover shocks are uncommon in original equipment designs for vehicles, though they have become widely available as aftermarket add-ons. Coilover shocks for cars have been considered specialty items for high performance and racing applications where they allow for significant reductions in overall vehicle height, and though high-quality aftermarket options with wide sturdy springs may provide improvements in vehicle performance, there is dispute over whether or not most aftermarket coilover shocks confer any material benefits to most drivers and may in fact reduce performance over original equipment installations.

Mono-tube

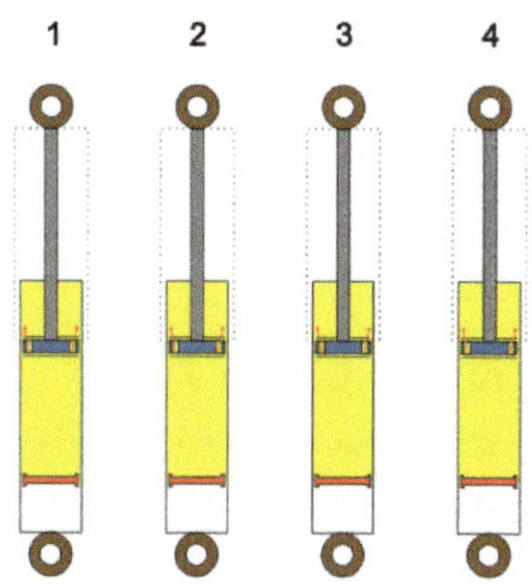

Hydraulic shock absorber monotube in different operational situations:
1) Drive slow or adjustments open
2) Like "1", but extension immediately after the compression
3) Drive fast adjustments or closed, you can see the bubbles of depression, which can lead to the phenomenon of cavitation
4) Like "3", but the extension immediately after the compression
Note: The volume change caused by the stem is considered.

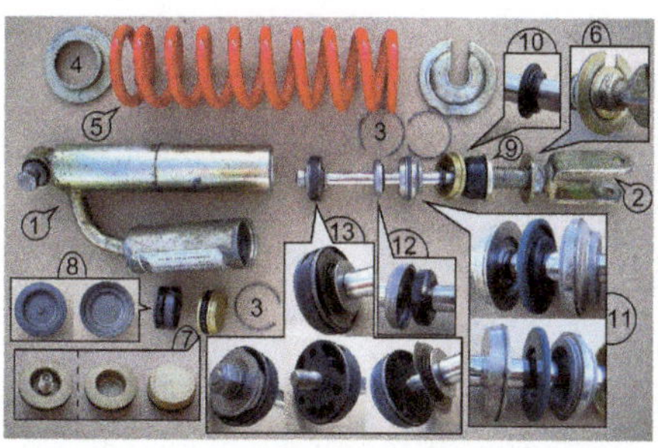

Absorber with gas tank connected rigidly, compared to most shock absorbers. It uses a diaphragm instead of a membrane, and does not contain a control valve for expansion of the pneumatic chamber.
Description:
1) Sheath and gas tank
2) Stem
3) Snap rings
4) Plate bearing spring
5) Spring
6) End cap and preload adjustment
7) Cap gas, present in versions both with or without gas valve (inverted profile)
8) Mobile diaphragm
9) Pad switch (compression)
10) Wiper
11) Oil seal assembly, and shock seal
12) Negative buffer pad or limit switch (extension)
13) Piston with sliding blades and seal

The principal design alternative to the twin-tube form has been the mono-tube shock absorber which was considered a revolutionary advancement when it appeared in the 1950s. As its name implies, the mono-tube shock, which is also a gas-pressurized shock and also comes in a coilover format, consists of only one tube, the pressure tube, though it has two pistons. These pistons are called the working piston and the dividing or floating piston, and they move in relative synchrony inside the pressure tube in response to changes in road smoothness. The two pistons also completely separate the shock's fluid and gas components. The mono-tube shock absorber is consistently a much longer overall design than the twin-tubes, making it difficult to mount in passenger cars designed for twin-tube shocks. However, unlike the twin-tubes, the mono-tube shock can be mounted either way— it does not have any directionality. It also does not have a compression valve, whose role has been taken up by the dividing piston, and although it contains nitrogen gas, the gas in a mono-tube shock is under *high* pressure (260-360 p.s.i. or so) which can actually help it to support some of the vehicle's weight, something which no other shock absorber is designed to do.

Mercedes became the first auto manufacturer to install mono-tube shocks as standard equipment on some of their cars starting in 1958. They were manufactured by Bilstein, patented the design and first appeared in 1954s. Because the design was patented, no other manufacturer could use it until 1971 when the patent expired.

Theoretical Approaches

There are several commonly used principles behind shock absorption:

- Hysteresis of structural material, for example the compression of rubber disks, stretching of rubber bands and cords, bending of steel springs, or twisting of torsion bars. Hysteresis is the tendency for otherwise elastic materials to rebound with less force than was required to deform them. Simple vehicles with no separate shock absorbers are damped, to some extent, by the hysteresis of their springs and frames.

- Dry friction as used in wheel brakes, by using disks (classically made of leather) at the pivot of a lever, with friction forced by springs. Used in early automobiles such as the Ford Model T, up through some British cars of the 1940s. Although now considered obsolete, an advantage of this system is its mechanical simplicity; the degree of damping can be easily adjusted by tightening or loosening the screw clamping the disks, and it can be easily rebuilt with simple hand tools. A disadvantage is that the damping force tends not to increase with the speed of the vertical motion.

- Solid state, tapered chain shock absorbers, using one or more tapered, axial alignment(s) of granular spheres, typically made of metals such as nitinol, in a casing.

- Fluid friction, for example the flow of fluid through a narrow orifice (hydraulics), constitutes the vast majority of automotive shock absorbers. This design first appeared on Mors racing cars in 1902. One advantage of this type is, by using special internal valving, the absorber may be made relatively soft to compression (allowing a soft response to a bump) and relatively stiff to extension, controlling "rebound", which is the vehicle response to energy stored in the springs; similarly, a series of valves controlled by springs can change the degree of stiffness according to the velocity of the impact or rebound. Specialized shock absorbers for racing purposes may allow the front end of a dragster to rise with minimal resistance under acceleration, then strongly resist letting it settle, thereby maintaining a desirable rearward weight distribution for enhanced traction.

- Compression of a gas, for example pneumatic shock absorbers, which can act like springs as the air pressure is building to resist the force on it. Enclosed gas is compressible, so equipment is less subject to shock damage. This concept was first applied in series production on Citroën cars in 1954. Today, many shock absorbers are pressurized with compressed nitrogen, to reduce the tendency for the oil to cavitate under heavy use. This causes foaming which temporarily reduces the damping ability of the unit. In very heavy duty units used for racing or off-road use, there may even be a secondary cylinder connected to the shock absorber to act as a reservoir for the oil and pressurized gas. In aircraft landing gear, air shock absorbers may be combined with hydraulic damping to reduce bounce. Such struts are called *oleo struts* (combining oil and air) .
- Inertial resistance to acceleration, for example prior to 1966 , the Citroën 2CV had shock absorbers that damp wheel bounce with no external moving parts. These consisted of a spring-mounted 3.5 kg (7.75 lb) iron weight inside a vertical cylinder and are similar to, yet much smaller than versions of the tuned mass dampers used on tall buildings.
- Composite hydropneumatic suspension combines many suspension elements in a single device: spring action, shock absorption, ride-height control, and self leveling suspension. This combines the advantages of gas compressibility and the ability of hydraulic machinery to apply force multiplication.
- Conventional shock absorbers can be combined with air suspension springs - an alternate way to achieve ride-height control, and self leveling suspension.
- In an electrorheological fluid damper, an electric field changes the viscosity of the oil. This principle allows semi-active damper applications in automotive and various industries.
- Magnetic field variation: a magneto rheological damper changes its fluid characteristics through an electromagnet.
- The effect of a shock absorber at high (sound) frequencies is usually limited by using a compressible gas as the working fluid or mounting it with rubber bushings.

Special Features

- Some shock absorbers allow tuning of the ride via control of the valve by a manual adjustment provided at the shock absorber.
- In more expensive vehicles the valves may be remotely adjustable, offering the driver control of the ride at will while the vehicle is operated.
- Additional control can be provided by dynamic valve control via computer in response to sensors, giving both a smooth ride and a firm suspension when needed, allowing ride height adjustment or even ride height control.

- Ride height control is especially desirable in highway vehicles intended for occasional rough road use, as a means of improving handling and reducing aerodynamic drag by lowering the vehicle when operating on improved high speed roads.

Motorcycle Frame

Triton: A Triumph engine in a tubular steel Norton Featherbed frame

A motorcycle frame is a motorcycle's core structure. It supports the engine, provides the hinge points for both front and rear suspension, and supports the rider and any pillion or luggage. Also attached to the frame are the fuel tank and battery. At the front of the frame is found the head tube that holds the pivoting front fork, while at the rear there is a pivot point for the swingarm. Some motorcycles include the engine as a load-bearing stressed member; while some others bikes do not use a single frame, but instead have a front and a rear subframe attached to the engine.

Materials

In the early days, motorcycles were little more than motorised bicycles, and consequently frames were tubular steel. While the use of steel tubing is still common, in modern times other materials such as titanium, aluminium, magnesium, and carbon-fibre, along with composites of these materials, are now used. As different motorcycles have varying design parameters (such as cost, complexity, weight distribution, stiffness, power output and speed), there is no single ideal frame design, and designers must make an informed decision of the optimum choice.

Steel

In Europe and the USA, steel tubing was the default material until recent times. All the major manufacturers (AJS, Ariel, BSA, Matchless, Norton, Sunbeam, Triumph, Velocette, BMW, DKW, Ducati, Moto-Guzzi, Harley-Davidson and Indian) used steel tubing.

Examples

- Norton Featherbed frame
- Early Honda CBR600
- Most Ducati Motorcycles

Aluminium

Examples

- Honda VFR750
- Yamaha YZF-R1

Carbon Fibre

Examples

- Ducati Desmosedici
- Ducati 1299 Superleggera
- MotoCzysz C1

Magnesium

Magnesium framed Elf 5

Examples

- 1988 Elf ROC-Honda Elf5-NSR500 500 cc Grand Prix

Titanium

Examples

- 1971 Titanium Husqvarna Inter-AMA Motocross

Composite

Greeves 250DCX Sportsman 1962

Examples

- Bimota SB8K, composed of two aluminium alloy beams and carbon fibre plates
- MV Agusta F4 750 Serie Oro, magnesium and aluminium
- Greeves 250DCX Sportsman, with cast alloy "downtube", tubular steel rear sub-frame and semi-monocoque spine.

Types

Spine or Backbone

The motorcycle engine is suspended from a single spine.

Examples

- Honda CB92 Benly
- MZ TS250

Single Cradle

The motorcycle engine is held in a single cradle with a single spine.

Examples

- Honda CG125

Half-duplex Cradle

The motorcycle engine is held in a double cradle with a single spine and single down-tube.

Examples

- Suzuki GSX250
- Suzuki TS50ER

Double cradle frame on Honda CB750

Full Duplex Cradle

The motorcycle engine is held in place within a pair of separate cradles. The Norton Featherbed frame was the classic example, but many "duplex" frames actually have a single spine beneath the tank.

Examples

- Norton Manx
- Suzuki TS50X

Twin-beam Frame on Kawasaki ZX-10R

Aluminum beam frame on Buell Lightning

Perimeter

Also called Beam or twin spar, two beams wrap around the engine to join the steering head and swing arm in the shortest distance possible for better rigidity. Beams are usually made of pressed metal (steel/aluminium). The trellis frame employs the same concept but uses welded members to form a trellis instead of pressed metal.

Examples

- 1885 Daimler *Reitwagen*
- Honda CBR1000RR
- Yamaha FZR600

Pressed

Pressed steel frame on Ariel Arrow

The frame is pressed or stamped from sheet metal to form a car-type semi-monocoque. The frame may be entirely pressed (Ariel Arrow), or may have just a pressed aft section connected to the steering head by a conventional steel tubular spine (Honda Super Cub). Both the Super Cub and the Arrow also have pressed steel forks, instead of conventional telescopic forks.

Examples

- Ariel Arrow
- Honda Super Cub

Monocoque

More common in the car world, a monocoque frame comprises a structure where loads are supported through its external skin. On bikes they are used almost exclusively on racing motorcycles.

2006 ZX-14 cutaway showing its aluminium monocoque frame

Examples

- 2000 Kawasaki Ninja ZX-12R
- 2006 Kawasaki Ninja ZX-14
- 2012 Ducati 1199

Semi-monocoque

JPS Norton with semi-monocoque frame

If a "monocoque" frame uses additional support, such as tubes or longerons, the frame is more properly called a "semi-monocoque". An example is Peter Williams' semi-monocoque Norton racer.

Trellis

Steel trellis frame (red) on a Ducati Monster 1000. The engine is a stressed member.

A trellis frame connects the steering head to the swingarm pivot as directly as possible using metal tube arranged in triangulated reinforcement. Using lattice girder principles, a trellis frame is typically constructed of round or oval section metal tubular segments that are welded or brazed together. A well-designed trellis frame should provide a strong, lightweight structure that simplifies placement of engine & components, and gives good access for maintenance. Although construction of a trellis frame needs a more complicated process than, say, an alloy beam frame, it requires only a simple jig and a competent welder. No heavy capital outlay is required, so a trellis frame is ideal for a model that may be made in relatively small numbers. For this reason, the trellis frame has found favour with European manufacturers, and Ducati in particular.

Some motorcycles, such as the Yamaha TRX850, have hybrid frames that employ alloy castings at the swingarm pivot area. Another variation is to suspend the engine from a trellis frame, but have the swingarm pivot cast into the rear of the engine.

Examples

- Most modern Ducatis
- KTM 690 Enduro
- Honda VTR250
- Suzuki Bandit 400 1989 Model

Engine as a Stressed Member

For rider comfort, a motorcycle's engine can be mounted on rubber bushings to isolate vibration from the rest of the machine. This strategy means the engine contributes little to frame stiffness, and absorbing rather than dissipating vibration can lead to stress damage to the frame, exhaust pipes, and other parts.

Harley-Davidson Model W with structural tubes bolted directly to engine case to complete the frame triangle

Instead, if the engine is rigidly mounted to the frame, vibrations pass to and are dissipated via the whole frame, and the rider. Rigid mounting allows the engine to contribute to the overall stiffness of the frame. It also becomes possible to mount the swingarm directly to the engine rather than the frame, avoiding the need for frame members extending downward to the swingarm pivot. By increasing the number of mounting points between the engine and frame, vibrations and stress can be better dissipated in the frame, typically creating a triangle between the swingarm in the rear, the cylinder head at the top and the lower crankcase area at the front. If a rigidly mounted engine not only contributes to, but is critical to, the stiffness of the frame, and is an integral part of closing the triangle or trellis structure that transfers force from the headstock to the swingarm, to the point that without the engine the frame would be deformed, the engine is called a stressed member, or a lifted engine. Sharing the load between the engine and frame reduces the overall weight of the motorcycle.

Stressed member engines were pioneered at least as early as the 1916 Harley-Davidson 8-valve racer, and incorporated in the production Harley-Davidson Model W by 1919. This was called a keystone, or diamond, frame. The 1946 Vincent Series B Rapide was designed with an advanced chassis, termed a “tour de force for its day,” that included a stressed member engine. During early testing of the 1983 Kawasaki GPZ900R, twin downtubes were included, creating a full cradle, but the downtubes were found to carry little load, so they were removed, relying entirely on the combination of the steel backbone and engine for chassis rigidity. BMW’s R1100 series twins of 1994 relieved the frame of stress entirely, with the engine carrying the total load from the front Telelever fork to the rear Monolever.

Stiffness

Frame stiffness is a problem for motorcycle designers, as it was for the bicycle frames that motorcycles are descendants of.

Modifying the stiffness of a factory-produced frame can be undertaken to improve handling characteristics. This is often done by triangulating the factory frame. Triangulation is a technique used in many engineering applications to stiffen structures. However doing so can also have undesirable effects if it overloads other parts of the frame, as a flexible frame acts as a spring to absorb some loads.

Lateral Stiffness

In the 21st century, advances like high power engines driving high traction tires, and better-performing suspension components, especially forks, led to a situation where designs with increased overall frame stiffness were made available to consumers. Analysts differ on whether infinite lateral stiffness is desirable, or whether a finite degree of built-in flex is preferable.[

Featherbed Frame

The featherbed frame was a motorcycle frame developed by the British Norton motorcycle company to improve the performance of their racing motorcycles around the twisting and demanding Isle of Man TT course in 1950. It was considered revolutionary at the time, and the best handling frame that a racer could have. Later adopted for Norton production motorcycles, it was also widely used by builders of custom hybrids such as the Triton, becoming legendary and remaining influential to this day.

1990s replica Norton with replica Featherbed frame

The Featherbed inspired other frame builders who based their own products on similar principles, including the heavyweight Münch Mammoth and a lightweight version for a 250 cc BSA C15 engine.

The Featherbed was replaced by the Norton Isolastic frame in 1967 for the then newly-developed Norton Commando which used a rubber-mounted engine and gearbox, although Norton continued to offer the *Mercury* with a Featherbed frame until production ceased by 1970. Replicas of the frame continue to be produced by specialist companies.

Triton. A Triumph 650 cc pre-unit engine and gearbox in a wideline Norton Featherbed frame

In 1949 brothers Rex and Cromie McCandless offered Norton a new frame to support their successful 500cc race single. Rex McCandless was a self-taught Belfast motorcycle engineer and raced competitively with his brother on a Triumph Tiger 100. He had made several improvements to the Triumph, notably an innovative new frame with a swinging arm fitted with vertical hydraulic shock absorbers from a Citroen car. BSA bought several of his converted motorcycles but Norton saw the real opportunity and contracted him to work exclusively for them from 1949. The Norton Motorcycle Company were concerned at the reliability of their plunger (or "Garden Gate") frame, as several had broken through the stress of racing. Norton engineer Joe Craig solved the problems by making the frames heavier but handling suffered as a consequence.

Norton commissioned the McCandless brothers to design a complete frame, incorporating a swinging arm. McCandless' finished design was expensive, as it required over forty feet of the best Reynolds steel tubing. It was a welded twin loop with a swinging arm fitted with their own design of shock absorbers, with a heavily braced cross-over headstock. In two months a prototype motorcycle with the new frame was on the test track and it was tested on the Isle of Man in the winter of 1949. It performed well and Norton decided that the Norton works team would have motorcycles with the new frames. The Norton works was not well equipped so the sif-bronze welding was undertaken by the McCandless brothers who produced the eight frames for the racing team by hand.

The Patent

Norton applied for a patent for the design on 13 October 1949 and it was granted as reference 664,667 but the completed specification was not published until 9 January 1952. The Featherbed frame was simply constructed.

This invention relates to a new or improved frame for a motorcycle which comprises two substantially parallel rectangular loops each formed from a single length of tubing, and the ends of the tube forming each loop cross and are welded to each other at the top front corner of the loop, the free ends of the tube which extend beyond the crossing point being welded to the side of an inclined head tube adjacent to the top and bottom thereof. The assembled frame is extremely strong for its weight and designed to provide the maximum resistance to any stresses applied to the frame by road shocks or by the driving torque of the power unit.

The Featherbed Name

Harold Daniell was a successful Isle of Man TT racer with three victories and several placings in the Tourist Trophy races and the Manx Grand Prix. After testing the new Norton frame in 1950 he declared that it was like "riding on a featherbed" compared with riding the "garden gate" — and it has been called the featherbed frame ever since.

The term feather bed was used in the 1933 to 1938 Brough Superior catalogs in their press section. As printed in the 1933 Brough Superior catalog:

The "Brough Superior" Rear Spring Frame, to quote "Castor" of "Motor Cycling," renders "pitching or wobble non-existent, impossible. A feather bed could scarcely be safer" -- and this in relating his experience on a Special "Brough Superior" S.S. 100, on which he did 106 m.p.h. in second gear on the road !

Racing Success

Further testing took place at Montlhery race track with four motorcycles running flat-out for two days. The new frame stood up well to tests and saw its UK launch at Blandford Camp in Dorset in April 1950. Geoff Duke had won the Senior Clubmans TT and the Senior Manx Grand Prix in 1949 on the earlier type Nortons, so was a clear choice for Norton to really put the new Featherbed-framed race bikes to the test. Duke won the race on the new design and several racing successes followed with Norton winning first three places in the 1950 Senior and Junior TT's. In the Friday Senior TT Duke set a new lap record of 93.33mph and also broke the overall race record, finishing in two hours, 51 minutes and 45 seconds; he had previously finished second to Artie Bell (Norton) in Monday's Junior TT. (Harold Daniell's Norton was third.) When it came to the bends on the twisting Island course the new frame gave the Nortons a distinct advantage.

Featherbed frames were also successfully modified for use in off road racing or motorcycle scrambling.

Featherbed Variants

Manx Racer in Reynolds 531

Weight and strength were key factors in the design of the featherbed frame for the Norton racing team's Manx. 16- Gauge Reynolds 531, a high-tensile manganese-molybdenum steel alloy, was used as it allowed the frame tubes to be made thinner for the same strength, as well as making for a more responsive frame. All the joints were Sifbronze welded, – a relatively low-temperature flame-braze – except for the sub frame which was initially bolted-on but welded in later versions.

Norton International

In 1953 the Norton International was relaunched with a new version of the Featherbed frame made from grade A mild steel.

Single and Twin Cylinder Roadsters in Featherbeds

Having earlier installed the Dominator twin-cylinder engine of 500 cc and 600 cc, in 1959 Norton put the old single cylinder Model 50 (350 cc) and the ES2 (500 cc) into the Featherbed frame to rationalise production. Using grade A mild steel, the size of this engine determined the space between the top and bottom rails of the full duplex cradle. In 1960 the top rails were installed at the rear of the tank. Riders complained that these *wideline* Featherbed frames were uncomfortably wide at 11.5 inches (29 cm) but it was not until 1960 that the top runs of the frame were narrowed towards the front of the seat, with corresponding overall styling changes including tank and seat to create the *slimline* frame.

The slimline was used until the last of the vertical twin cylinder models in the late 1960s, the Norton Mercury, a limited-production run of single carburettor 650 cc machines based on the Dominator; the Norton Commando with its new frame design and angled-forward engine having been launched at the Earls Court show in 1967 took over as the 750 cc range-topper, later enlarged to 830 cc, but badged as '850'.

Motorcycle Fairing

Kawasaki Ninja ZX-12R full fairing

A motorcycle fairing is a shell placed over the frame of some motorcycles, especially racing motorcycles and sport bikes, with the primary purpose to reduce air drag. The secondary functions are the protection of the rider from airborne hazards and wind-induced hypothermia and of the engine components in the case of an accident. A motorcycle windshield will almost always be integrated into the design of the fairing.

The major benefit of a fairing on sport touring and touring motorcycles is a reduction in aerodynamic drag, which allows for reduced fuel consumption and permits higher speeds at lower engine rpm, which in turn increases engine life.

A motorcycle may have a front fairing, a rear fairing, a belly fairing, or any combination of these. Alternatively, a single fairing may partially or fully enclose the entire motorcycle, and may even enclose the rider.

History

Royal Enfield Bullet with factory-optional *Airflow* touring-fairing and matching front wheel streamlining

The importance of streamlining was known very early in the 20th century, and some streamlining was seen on racing motorcycles as early as the 1920s. Although motorcycles generally have a much higher power-to-weight ratio than cars, bikes – and particularly the rider – are much less streamlined and the effects of aerodynamic drag on motorcycles are very significant. Consequently, any reduction in a motorcycle's drag coefficient pays dividends in improved performance.

The term fairing came into use in aircraft aerodynamics with regard to smoothing airflow over a juncture of components where airflow was disrupted. Early streamlining was often unsuccessful resulting in instability. Handlebar fairings, such as those on Harley-Davidson Tourers, sometimes upset the balance of a motorcycle, inducing wobble.

Originally the fairings were cowlings put around the front of the bike, increasing its frontal area. Gradually they became an integral part of the design. Modern fairings increase the frontal area at most by 5% compared to a naked machine. Fairings may carry headlights, instruments, and other items. If the fairing is mounted on the frame, placing other equipment on the fairing reduces the weight and rotational inertia of the steering assembly, improving the handling.

The BMW R100RS, produced from 1976 to 1984, was the first mass-market sport touring motorcycle to be offered with a full fairing as standard, and marked the beginning of wider adoption of fairings on sports and touring types of motorcycles. The integrated design included a development of the frame-mounted tail fairing at the rear of the removable dual seat accessing a storage compartment used on the BMW R90S from 1973, also being the first example of a factory-fitted head (or nose) fairing.

Types

Front Fairings

Dustbin Fairing

NSU Sportmax 1955, dustbin fairing

JPS Norton with handlebar blisters

A single piece, streamlined shell covering the front half of a motorcycle resembling the nose of an aircraft, sometimes referred to as torpedo fairing. It dramatically reduced the frontal drag, but it was banned by Fédération Internationale de Motocyclisme (FIM) from racing in 1958, because it was thought that the frontal point of wind pressure made them highly unstable even with small amounts of yaw. Other reasons cited for the ban were to ensure adequate steering range (lock-to-lock) and stability in crosswinds. FIM regulations forbid streamlining beyond the wheel spindles and require the rider's arms and legs to be visible from the side. However, Peter Williams was permitted to give his 1973/74 JPS Norton a Peel-type fairing incorporating handlebar blisters which helped to reduce the drag coefficient to 0.39.

Dolphin Fairing

NSU Rennmax 250 cc Delphin, the original instance of the name *Dolphin*

This was called so because, in early models, the front wheel mudguard streamlined with the rising windshield part of the fairing resembled the dolphin's snout from the side view. Further developments on this design became the norm after dustbin fairings were banned.

Full fairing

A full fairing is a large front-mounted fairing, and should not be confused with cabin motorcycle or streamliner motorcycle fairings which fully or partially enclose the entire motorcycle.

Full fairings cover both upper and lower portions of the motorcycle, as distinct from a half fairing, which only has an upper section, and leaves the lower half of the motorcycle exposed. The fairing on a race or sport bike is meant as an aerodynamic aid, so the windscreen is rarely looked through. If the rider is sitting up at speed he will be buffeted by his rapid progress through the air and act as a parachute, slowing the bike, while if the rider lies flat on the tank behind the windscreen he generates much less aerodynamic drag. The high windscreen and handlebar width of a touring fairing protect the upright rider from the worst of this, and the windscreen is functional. Full fairings can also provide protection to the engine and chassis in the event of a crash where the fairings, rather than the engine covers and/or frame, slide on the road.

Half Fairing

Suzuki SV650s with OEM half-fairing & rear-fairing, plus aftermarket bellypan

Half fairings usually feature a windscreen and extend below the handlebars, possibly as far down as the sides of the cylinder block, but generally do not cover the sides of the crankcase or gearbox. Aftermarket kits – 'lowers' – are available to extend some half fairings into full fairings. Due to the popularity of these kits, some motorcycle manufacturers have started to supply their own full fairing conversion kits and even offer their half faired models new with a full factory-fitted kit.

Quarter Fairing

A windscreen and minimal fairing extending around the headlamp fixed to the triple clamp. Also called a café fairing or bikini fairing, it stops well below the level of the rider's head, relying primarily on air deflection to protect the rider's head and chest from the slipstream.

BMW R90S quarter fairing

Belly Pan

Quarter and half fairings are often paired with a belly pan below the engine for diverting air flow away from under the engine to reduce aerodynamic lift, as well as cosmetic reasons. Some track day or racing rules require belly pans to catch leaked fluids.

Rear Fairings

Tail fairing on a Kawasaki S2, partly enclosing the tail light

A tail fairing, sometimes referred to as a tail unit or race tail, is mounted behind the seat and rider. Some also extend to the sides of, and below, the saddle and usually providing a small storage space.

Kawasaki became the first manufacturer to incorporate a tail fairing on a mass production motorcycle when tail units were fitted to the 1972 250cc S1 Mach I, 350cc S2 Mach II, and 750cc H2 Mach IV models.

Streamliner

A streamliner motorcycle, such as the NSU Delphin III, uses a fairing which completely encloses both the entire motorcycle and the rider to provide the best possible drag coefficient ratio.

Materials

Acrylonitrile butadiene styrene (ABS) plastic is commonly used in original equipment sport bikes and certain aftermarket fairing manufacturers due to its strong, flexible and light weight properties. The advantage of ABS over other plastics is that it combines the strength and rigidity of acrylonitrile and styrene with the toughness of polybutadiene rubber. The proportions of each property vary based on the targeted result.

There are two common methods of producing an ABS plastic fairing: injection and compression molding.

- Injection molding: ABS plastic is melted and injected into mold cavity. Constant pressure is applied to allow for material shrinkage. The plastic then cools and hardens in the mold. Injection molds allows for uniform thickness throughout the entire piece. It gives the most accurate end product that fits well. It is how most manufacturers make their plastics.

- Compression molding: The plastic is generally preheated is placed into a heated metal mold cavity and pressure is applied to force the plastic to contract and take the shape of the mold. Heat and pressure is kept until the plastic cures the mold. The excess plastic is cut away and removed from the mold. Disadvantages to compression mold include varying product consistency and flashing, which is excess material attached to the molded part that needs to be removed where two or more parts of the mold meet.

Fiberglass is made of woven fibers, and is used as a reinforcing agent for many polymer products. The composite properly known as glass-reinforced plastic (GRP), is normally referred to by the name of its reinforcing material. Fiberglass fairings are commonly used on the race track. In most cases fiberglass is lighter, and more durable than ABS Plastic. Damaged fiberglass can be repaired by applying new layers of woven fiberglass cloth mixed with a polymer such as epoxy over the damaged area, followed by sanding and finishing.

Carbon-fiber-reinforced polymer is the lightest, but most expensive, fairing material. It is used on the most extreme sport and racing motorcycle fairings.

References

- Cossalter, V. "BMW Telelever and telescopic forks: which is better ?". Archived from the original on 2010-05-16. Retrieved 2012-04-07
- Wilson, David Gordon; Jim Papadopoulos (2004). Bicycling Science (Third ed.). The MIT Press. pp. 263–390. ISBN 0-262-73154-1
- Sharp, R. S. (July 2007). "Motorcycle Steering Control by Road Preview". Journal of Dynamic Systems, Measurement, and Control. ASME. 129 (July 2007): 373–381. doi:10.1115/1.2745842
- Jones, David E. H. (1970). "The stability of the bicycle" (PDF). Physics Today. 23 (4): 34–40. Bibcode:1970PhT....23d..34J. doi:10.1063/1.3022064. Retrieved 2008-09-09
- Sharp, R.S. (1985). "The Lateral Dynamics of Motorcycles and Bicycles". Vehicle System Dynamics. 14 (4–6): 265–283. doi:10.1080/00423118508968834
- Whitt, Frank R.; David G. Wilson (1982). Bicycling Science (Second ed.). Massachusetts Institute of Technology. pp. 188, 198–233. ISBN 0-262-23111-5
- Sharp, R. S. (1971). "The stability and control of motorcycles". Journal of Mechanical Engineering Science. 13 (5): 316–329. doi:10.1243/JMES_JOUR_1971_013_051_02
- Zinn, Lennard (2004-12-21). "Technical Q&A with Lennard Zinn — Rake, trail, offset". Velo News. Archived from the original on 2006-06-19. Retrieved 2006-08-04
- Foale, Tony (2006). Motorcycle Handling and Chassis Design (Second ed.). Tony Foale Designs. ISBN 978-84-933286-3-4
- Roe, G. E. and Thorpe, T. E. "A Solution of the low-speed wheel flutter instability in motorcycles" Journal Mechanical Engineering Science V 18 No. 2 1976

- Brown, Sheldon (2006). “Sheldon Brown’s Bicycle Glossary”. Sheldon Brown. Archived from the original on 13 August 2006. Retrieved 2006-08-08
- Manfred Plochl; Johannes Edelmann; Bernhard Angrosch & Christoph Ott (July 2011). “On the wobble mode of a bicycle”. Vehicle System Dynamics. 50 (3): 415–429. doi:10.1080/00423114.2011.594164
- Wilson, David Gordon (2004), Bicycling Science (3rd ed.), Massachusetts Institute of Technology, p. 245, ISBN 978-0-262-23237-1
- Kurtus, Ron (2005-11-02). “Coefficient of Friction Values for Clean Surfaces”. Archived from the original on 2007-09-29. Retrieved 2006-08-07
- Wilson, Hugo (1995). “The A-Z of Motorcycles”. The Encyclopedia of the Motorcycle. London: Dorling Kindersley. p. 100. ISBN 0-7513-0206-6
- Cleary and Mohazzabi (15 July 2011). “On the stability of a bicycle on rollers”. European Journal of Physics. Retrieved 2012-11-08
- Biral, F; Bortoluzzi D; Cossalter V; Da Lio M (2003). “Experimental study of motorcycle transfer functions for evaluating handling”. Vehicle System Dynamics. 39 (1): 1–25. doi:10.1076/vesd.39.1.1.8243
- Jack Korpela (13 August 2012). “Spiderman rained out in Norwalk 6.01 and successful parachute test”. Cycle Drag. Retrieved 2015-06-23

4

Various Automotive Styling Features

Custom cars are cars that are made as per the requirements and choice of the driver. The machine is altered in order to improve its performance and to make it look different. The features added are paint, engines swapping, etc. This chapter is a compilation of the various automotive styling features that form an integral part of the broader subject matter.

Custom Car

A custom car is a passenger vehicle that has been substantially modified in either of the following two ways:

1. a custom car may be altered to improve its performance, often by altering or replacing the engine and transmission
2. a custom car may be a personal "styling" statement, making the car look unlike any car as delivered from the factory.

Although the two are related, custom cars are distinct from hot rods. The extent of this difference has been the subject of debate among customizers and rodders for decades. Additionally, a street rod can be considered a custom.

'32 3-window with a classic-style flame job and Moon tank, reminiscent of Chapouris' *California Kid*.

Custom '51 Merc with red "ghost flames" and Appletons

A pre-war custom car by Coachcraft

The iconic "T-bucket" custom.

Front suspension of yellow lowboy Deuce roadster. Note color-matched springs on coilover shocks, tube axle, vented disc brakes.

'41 Willys. Note the non-stock one-piece windshield.

'28 A roadster with Kelsey-Hayes wire wheels

History

A development of hot rodding, the change in name corresponded to the change in the design of the cars being modified. The first hot rods were pre-World War II cars, with running boards and simple fenders over the wheels. Early model cars (1929 to 1934) were modified by removing the running boards and either removing the fenders entirely or replacing them with very light cycle fenders. Later models usually had fender skirts installed. The "gow job" morphed into the hot rod in the early to middle 1950s.

Typical of builds from before World War Two were '35 Ford wire-spoke wheels.

Many cars were "hopped up" with engine modifications such as adding additional carburetors, high compression heads, and dual exhausts. Engine swaps were often done, the object of which was to put the most powerful engine in the lightest possible frame and body combination.

The suspension was usually altered. Initially this involved lowering the rear end as much as possible with the use of lowering blocks on the rear springs. Later cars were given a rake job either adding a dropped front axle or heating front coil springs to make the front end of the car much lower than the rear. Immediately postwar, most rods would change from mechanical to hydraulic ("juice") brakes and from bulb to sealed-beam headlights.

The mid-1950s and early 1960s custom Deuce was typically fenderless and steeply chopped, and almost all Ford (or Mercury, with the 239 cu in (3,920 cc) flatty, introduced in 1939); a Halibrand quick-change rearend was also typical, and an Edelbrock intake manifold or Harman and Collins ignition magneto would not be uncommon. Reproduction spindles, brake drums, and backing based on the 1937s remain available today. Aftermarket flatty heads were available from Barney Navarro, Vic Edelbrock, and Offenhauser. The first intake manifold Edelbrock sold was a "slingshot" design for the flatty. Front suspension hairpins were adapted from sprint cars, such as the Kurtis Krafts. The first Jimmy supercharger on a V8 may have been by Navarro in 1950. Much later, rods and customs swapped the old solid rear axle for an independent rear, often from Jaguar. Sometimes the grille of one make of car replaced by another; the 1937 Buick grille was often used on a Ford. In the 1950s and 1960s, the grille swap of choice was the 1953 De Soto.

The original hot rods were plainly painted like the Model A Fords from which they had been built up, and only slowly begun to take on colors, and eventually fancy orange-yellow flamed hoods or "candy-like" deep acrylic finishes in the various colors.

With the change in automobile design to encase the wheels in fenders and to extend the hood to the full width of the car, the former practices were no longer possible. In addition, there was tremendous automotive advertising and subsequent public interest in the new models in the 1950s. Hence custom cars came into existence, swapping headlamp rings, grilles, bumpers, chrome side strips, and tail lights, as well as frenching and tunnelling head- and taillights. The bodies of the cars were changed by cutting through the sheet metal, removing bits to make the car lower, welding it back together, and adding a lot of lead to make the resulting form smooth (hence the term lead sled; lead has since been replaced by Bondo). By this means, chopping made the roof lower; sectioning made the body thinner from top to bottom. Channeling was cutting notches in the floorpan where the body touches the frame to lower the whole body. Fins were often added from other cars, or made up from sheet steel. In the custom car culture, someone who merely changed the appearance without also substantially improving the performance was looked down upon.

More recently, *Juxtapoz Magazine*, founded by the artist Robert Williams, has covered Kustom Kulture art.

Customization Style

Custom cars are distinct from cars in stock condition. Builders may adopt the visual and performance characteristics of some relevant modification styles, and combine these as desired. There are now several different custom themes, including:

- Rat rod: imitates (or exaggerates) the "unfinished" and amateur-built appearance of hot rods of the 1930s, 1940s, and 1950s
- Restomod - restored and modernized. Safety and convenience upgrades, such

as disc brakes, AC, etc., but can include fuel injection and overdrive upgrades, etc. Externally might resemble a stock car with period correct mods rather than customs.

- Street machines: Typically American cars with large-displacement engines modified for speed and often appearance.
- Street rod - consist largely of period specific vehicles and components, or emmulate visual characteristics of cars through the'40s vintage. There is a great deal of overlap here with hot rods.
- Muscle Cars are the Cars Before the year 2000 that were Customized and Modified for the Best Performance
- Tweeners - early 1960s cars between the styles of 1950s and the muscle cars.
- Modern: The style of contemporary cars. Most often contemporary components and paint finishes of modern cars are used.
- NOTE: Types overlap and blend often impossible to classify.

Features

Paint

Paint was an important concern. Once bodywork was done, the cars were painted unusual colors. Transparent but wildly colored candy-apple paint, applied atop a metallic undercoat, and metalflake paint, with aluminum glitter within candy-apple paint, appeared in the 1960s. These took many coats to produce a brilliant effect – which in hot climates had a tendency to flake off. This process and style of paint job was invented by Joe Bailon, a customizer from Northern California.

Customizers also continued the habit of adding decorative paint after the main coat was finished, of flames extending rearward from the front wheels, scallops, and hand-painted pinstripes of a contrasting color. The base color, most often a single coat, would be expected to be of a simpler paint. Flame jobs later spread to the hood, encompassing the entire front end, and have progressed from traditional reds and yellows to blues and greens and body-color "ghost" flames. One particular style of flames, called "crab claw flames", which is still prevalent today, is attributed to Dean Jeffries.

Painting has become such a part of the custom car scene that now in many custom car competitions, awards for custom paint are as highly sought after as awards for the cars themselves.

Engine Swaps

Engine swaps have always been commonplace. Once, the flathead, or "flatty", was the preference, supplanted by the early hemi in the 1950s and 1960s. By the 1970s, the small-block Chevy was the most common option, and since the 1980s, the 350 cu in (5.7 l) Chevy has been almost ubiquitous. More recently, the 325 cu in (5.3 l) Chevrolet LS has begun replacing the 350. Flatheads and early hemis have not entirely disappeared, but ready availability, ease of maintenance, and low cost of parts make the Chevrolet V8, in particular the first and third generation small block, the most frequent engine of choice.

Once customizing post-war cars caught on, some of the practices were extended to pre-war cars, which would have been called fendered rods, with more body work done on them. An alternate rule for disambiguation developed: hot rods had the engine behind the front suspension, while customs had the engine over the front suspension. The clearest example of this is Fords prior to 1949 had Henry Ford's old transverse front suspension, while 1949 models had a more modern suspension with the engine moved forward. However, an American museum has what could be the first true custom, built by Cletus Clobes in 1932, among its exhibits.

With the coming of the muscle car, and further to the high-performance luxury car, customization declined. One place where it persisted was the U.S. Southwest, where lowriders were built similar in concept to the earlier customs, but of post-1950s cars.

As the supply of usable antique steel bodies has dried up, companies such Westcott's, Harwood, Gibbon Fiberglass and Speedway Motors have begun to fabricate new fiberglass copies, while Classic Manufacturing and Supply, for one example, has been making a variety of new steel bodies since the 1970s. California's "junker" (or "crusher") law, which pays a nominal sum to take "gross polluters" off the road, has been criticized by enthusiasts (and by SEMA) for accelerating this trend.

Starting in the 1950s, it became popular among customizers to display their vehicles at drive-in restaurants. Among the largest and longest lasting was Johnie's Broiler in Downey, California. The practice continues today, especially in Southern California.

Customizers

Examples of notable customizers include George Barris, Bill Cushenberry, the Alexander Brothers, the "legendary" Gil Ayala, Darryl Starbird, Roy Brizio, Troy Trepanier (of Rad Rides by Troy), Boyd Coddington, Darryl Hollenbeck (working out of at Vintage Color Studios; winner of the 2016 America's Most Beautiful Roadster (AMBR) trophy with a custom Deuce) Harry Westergaard, Dave Stuckey, Dean Jeffries, Barry Lobeck, Phil Cool (who won the 1978 AMBR trophy with a bright orange Deuce, cover car for the July 1978 issue of *Hot Rod*), Troy Ladd of Hollywood Hot Rods, Doane Spencer (builder of a 1940s Deuce considered the template for the hiboy), "Posie", Ron Clark and Bob Kaiser (of Clarkaiser Customs), Joe Bailon (inventor of candy apple paint), Gene Winfield, Rick Dore Joe Wilhelm, "Magoo", Chip Foose, and Pete Chapouris.

Others, such as Von Dutch, are best known as custom painters. Several customizers have become famous beyond the automobile community, including Barris, Jeffries, and Coddington, thanks to their proximity to Hollywood; Barris designed TV's Batmobile, while Chapouris built the flamed '34 three-window coupé in the eponymous telefilm "The California Kid". Another Barris creation, *Ala Kart* (a '29 Ford Model A roadster pickup), made numerous appearances in film (usually in the background of diner scenes and such), after taking two AMBR wins in a row.

Some customizers have become well-enough known to be referred to by given name alone. These include Boyd (Coddington), Pete (Chapouris), and Jake (Jim Jacobs).

Awards

The highest award for customizers is the AMBR (America's Most Beautiful Roadster) trophy, presented annually at the Grand National Roadster Show since 1948 (also known within the customizer community as the Oakland Roadster Show until it was moved to Southern California in 2003). This competition has produced famous, and radical, customs.

Another is the Ridler Award, presented at the Detroit Autorama since 1964 in honor of show promoter Don Ridler. With one of the most unusual of car show entry requirements, winners of the prestigious Ridler Award are selected as the most outstanding from among cars being shown for the first time. This prompts builders of many high-end roadsters to first enter the Autorama first and then the Grand National show in order to have the chance to win top honors at both shows. Few cars and owners can claim this achievement.

Notable Customs

Some customs gained attention for winning the AMBR trophy, or for their outlandish styling. Notable among these is *Silhouette* and Ed Roth's *Mysterion*. Some of these more unusual projects turned into Hot Wheels cars, among them *The Red Baron*.

Other custom cars became notable for appearances in film (such as *Ala Kart* {1958}, *The California Kid* three-window {1973}, or the yellow deuce from "American Graffiti" {1973}) or television (such as *The Monkeemobile*, the "Munsters" hearse, or, more recently, Boyd's full-custom *Tool Time* '34, or Don Thelan's '33 three-window, *Eliminator*, built for the ZZ Top video). Specialist vehicles, such as the T/A, KITT, from *Knight Rider*, are not usually considered customs, but movie or TV cars, because they retain a mostly stock exterior.

Still others exemplified a trend. One of these is the '51 Merc built by the Barris brothers for Bob Hirohata in 1953, known forever after as the Hirohata Merc. Even without an appearance in film ("Runnin' Wild"), it is iconic of 1950s customs, and of how to do a Merc right. The same year, Neil Emory and Clayton Jensen built *Polynesian* for Jack Stewart, starting with a '50 Olds sedan. *Polynesian* made the cover of *Hot Rod* in August, and saw 54 pages of construction details in *Motor Trend Custom Car Annual* in 1954.

Language

Certain linguistic conventions are followed among rodders and customizers:

- The model year is rarely given in full, except when it might be confused, so a 1934 model is a '34, while a 2005 might be an '05 or not.
    - A '32 is usually a Deuce and most often a roadster, unless coupé is specified, and almost always a Ford.
    - A 1955, 1956, or 1957 is usually a Chevrolet.
    - A 1955, 1956, or 1957 Chevrolet is often called a Tri-Five.
- A 3- or 5-window is usually a Ford, unless specified.
- A flatty is a flathead V8 (always Ford, unless specified); a late (or late model) flatty is probably a Merc.
- A hemi ("*hem* ee") is always a 426, unless displacement (331, 354, or 392) is specified; a 426 is a hemi, unless Wedge is specified.
- A 392 is an early hemi.
    - A 331 or 354 is known to be an (early) hemi, but rarely referred to as such
- A 270 "Jimmy" was a 270 cubic inch GMC truck engine often used to replace a smaller displacement Chevrolet six cylinder.
- Units are routinely dropped, unless they are unclear, so a 426 cubic inch (in^3) displacement engine is simply referred to as a 426, a 5-liter (litre) displacement engine is a 5.0 ("five point oh"), and a 600 cubic feet per minute (cfm) carburetor is a 600. Engine displacement can be described in cubic inches or liters (for example, a 5.7-liter engine is also known as a 350 {"three fifty"}); this frequently depends on which units the user is most comfortable or familiar with.

The "cutoff year" as originally promoted by the *National Street Rod Association* (NSRA) is 1949. Many custom car shows will only accept 1948 and earlier models as entries, and many custom car organizations will not admit later model cars or trucks (also with some imports - this has been a gray area of what's acceptable e.g. an aircooled VW Beetle, a Big Three product manufactured overseas e.g. a Ford Capri built in the UK or a General Motors - Holden's product, not to mention captives), and/or a vintage import automobile with an American driveline transplant) but this practice is subject to change. Modern day custom car shows which allow the inclusion of musclecars have used the 1972 model year as the cutoff since it is considered the end of the musclecar era prior to the introduction of the catalytic converter. The NSRA has announced that starting in 2011 it will switch to a shifting year method where any owner with a car 30 years or older will be allowed membership. So in 2011 the owner of a 1981 model year vehicle will qualify, then in 2012 the owner of a 1982 model year vehicle will quality, and so on. Additionally, the Goodguys car show organization has moved the year limit for its "rod" shows from 1949 to 1954 in recent years.

Types

- Hi-Riser (automobile)
- Hot rod
- Kustom (cars)
- Lead sled
- Lowrider
- Rat rod
- Rice burner
- Three window coupe

Common Terms

Some other common terms:

- 3 deuces – arrangement of three 2-barrel (twin-choke) carburetors; distinct from Six Pak and Pontiac and Olds Tri-Power (also 3x2 arrangements)
- 3-window – 2-door coupé; so named for having rear window and one door window on each side
- 3 on the tree – three-speed manual transmission operated by a steering column mounted shifter.
- 4 on the floor – four-speed manual transmission operated by a floor mounted shifter.
- 5-window – 2-door coupé; so named for having rear window plus one door window and one quarter window on each side
- 97s – Stromberg carburetors

- A-bone — Model A coupé
- Appletons (sometimes Appleton spots) — spotlights, mounted in the A-pillars, similar to those used by police cars. Decorative but used through the 1950s as driving lights because headlights were poor, roads were narrow with no white lines at the edges, and there were no speed limits on most highways. Not much traffic either, especially at night.
- Ardun - Ford flathead V8 hemi heads designed by Zora Arkus Duntov (later known as THE Corvette guy)
- Baby moons - chrome small smooth convex hubcaps covering the wheel lug area. Full moons covered the entire wheel.
- Baby hemi, baby elephant. Dimensionally small early 1950s Chrysler corporation small cubic inch hemi V8s, 241 to 325 cid. Often used in the 1950s in hot rods.
- Barn fresh, barn find— a newly discovered vehicle typically found in storage (i.e. a barn, etc.), either long forgotten or abandoned, still in its original condition from when it was first stored
- Blue oval — Ford product (for the Ford badge)
- Blue dots —
    - Pontiac tail lights
    - Any taillight equipped with a blue crystal to give it a "purple-ish" appearance when illuminated. Illegal in many states.
- Bondo — brand name for a body filler putty, often used as a generic term for any such product
- Bowtie (sometimes *Red Bowtie*) — Chevrolet product (for the badge; the red bowtie refers Chevrolet Motorsport's logo)
- Bugcatcher intake — large scoop intake protruding through hood opening, or on cars with no hood.
- Bullnosing — Replacing the hood ornament with a "bullnose" chrome strip or filling the mounting hole with lead.
- Cabriolet (or cabrio) — A vehicle with a removable or retractable cloth top, characterized by integrated door window frames and crank up glass.
- Cherry — like new
- Channeled or channeling - lowering a vehicle by cutting out the floor and mount-

ing the body lower on the frame rails.

- Chopped — removing a section, usually of the window posts, to lower the roofline of a vehicle.
- Cobra killers - decorative wheel centers that stick out excessively 3-5 inches and have flipper qualities for more visual attraction.
- Convertible — retractable top car with no integral door window frames like the cabriolet. Has roll up glass in doors as opposed to roadsters that do not.
- C.I.D. (sometimes Cubic Inches or Inches) — cubic inches displacement
- Crank — crankshaft
- Cutouts - stub exhaust pipes installed behind the front wheels that allow uncapping for noise and power. In the 1950s were home made from gas tank filler necks with gas caps and water pipes with screw on caps.
- Dagmars — large front bumper "bullets" the actress) younger folks call them Dollys
- Dollys - Dagmars, after the country singer, Dolly Parton.
- Decked — trunklid trim removed
- Deuce —
    - '32 Ford (most often a roadster); now commonly on A frame WRONG rails
        - rarely, 1932 model of any manufacturer.
    - Chevy II Nova
- Duvall windshield — a v-shaped windshield with a center post, as opposed to the typical stock straight-across type.
- Elephant — Chrysler 426 Hemi
- Fat-fender — 1934-48 (U.S.) car
- Flatty — flathead engine (usually refers to a Ford; when specified, the Mercury-built model)
- Fordillac ("*for* di lack") — Ford with transplanted Cadillac V8 engine
- Frenched —
    - Antenna sunken into the body or fender

    - Headlight slightly sunken into fender
    - Tail lights slightly sunken into body or fender
- Gennie — genuine
- Hairpins — radius rods
- Hiboy (or highboy) — fenderless, but not lowered Distinct from gasser.
- Hopped up - A stock engine modified to increase performance
- Humpback or hump. Late 1930s sedans with a prominent rear trunk. Opposite of slantback/ smoothback.
- Inches — CID
- Indian (also "Tin Indian") — Pontiac (for the grille badge)
- Jimmy (or Jimmy Six) — GMC straight 6
    - Any GMC product
- Juice brakes—hydraulic brakes
- Lead sled — customized vehicle where lead has been melted and adhered to a metal body to smooth its surface, as filler. (Lead has since been replaced by Bondo.)
- Lakes pipes — straight exhaust pipes that run along the lower edge of a rod, typically near the rocker panels, without mufflers. The name comes from their original use on cars used on dry lakes by land speed racers.
- Loboy (or low boy, lowboy) — fenderless and lowered
- Mag
    - magnesium wheel, or steel or aluminum copy resembling one such
    - magneto
- Mill — any internal combustion engine on such a vehicle
- Moons (or Moon discs; incorrectly, moon discs) — plain flat chrome or aluminum hubcaps, originally adopted by land speed racers. Smaller examples are "baby moons". Named for Dean Moon.
- Mouse — small-block Chevy
- Nailhead — an early Buick V8

- Nerf bars — bumper horns
- NOS — New Old Stock: original-manufactured part, never installed, often in original packaging.
- Nosed — hood trim removed
- Phaeton - 4 dr. ragtop "roadster" also called a touring
- Phantom — body style never built by the original manufacturer (a term also adopted by model kit builders)
- Pinched rails - when deuce frame rails are narrowed under a Model A, which has a narrower front body.
- QJ — Quadrajet (Rochester 4-barrel)
- Q-jet — Quadrajet
- Ragtop — convertible
- Rat —Chevrolet Big-block
- Repop — reproduction (not NOS)
- Resto — restoration, or restored
- Roadster - A two door with removable or retracting top. No roll up side glass like true convertibles, usually snap on ising glass.
- Rockcrusher — Muncie M22 4-speed transmission
- Rake job — car with suspension modified to lower the front end
- Rocket — Oldsmobile, in particular their early V8s. A reference to the marque's logo.
- Sabrinas (Britain) — bumper bullets, similar to Dagmars. Named after Italian Europop singer..
- SBC — Chevrolet small-block engine
- SBF — Small-block Ford, usually one of the Ford Windsor engines
- Sectioning — removing an entire horizontal section of the body or top to bottom. Shoebox — '49-'54 Ford or 1955-57 Chevrolet (for the slab-sided appearance)
- Skirts - Covers installed on the openings on rear fenders
- Slantback — sedan with forward-angled but straight rear window and sheet-

metal. Also referred to as slick back, slicky, smoothback, smoothy. Distinct from straightback.

- Smoothies - chrome steel wheels with no brake vent holes. Usually with baby moons or spiders.
- Sombreros — '47-'51 Cadillac hubcaps
- Souped (souped up) — hopped up, performance improved (more common in 1940s and 1950s)
- Spiders - a decrotive chrome insert covering the bearing grease cover and lugs nuts.
- Spinner knobs, suicide knobs - an egg sized knob mounted on the steering wheel to assist rapid turning. Many cars needed 5 or more turns of the wheel from full right to full left. Knobs often had pin-up girl pictures.
- Steelies — stock steel rims
- Stock — original equipment
- Stone stock — all-original (usually referring to a project's starting condition); unmodified ("'53 Merc with a stone stock 350").
- Stovebolt — Chevy straight 6
- Straightback — sedan with vertical rear window and sheetmetal. (Known as squareback in the VW community.)
- Street rod - A modified car licensed for use on streets and highways. --Traditionally designated as '48 and older by enthusiasts, SEMA, and states that issue specific street rod license plates. '49 and newer receive custom car plates. Antique, street rods, custom rod, and normal plates carry different rules under SEMA rules for state applied licenses.
- Studillac ("*stewed* i lack") — Studebaker with transplanted Cadillac V8 engine
- Suicided — changed from front- to rear-hinged ("suicide door")
- Suicide front end — a front axle configuration where it is mounted forward of the front cross member or the end of the frame rails.
- Suicide knob
- Taildragger - A vehicle lowered more in the rear than front. Often seen on leadsleds. Often a regionalized trend.
- Tin Indian — Pontiac (for the grille badge)

- Toploader — Ford 4-speed manual transmission
- Touring
- Track T — Model T roadster built in the style of a dirt track race car
- Trailer queen — pejorative term used in some circles for pure show cars which are never driven
- Tri-Five — 1955, 1956, or 1957 Chevrolet
- Tuck-and-roll — upholstery technique creating a "pleated" look
- Tunneled — deeply sunken into fender
- V-butted (or vee-butted) — with windshield center post deleted, original panes meeting in the middle (distinct from fitting a one-piece windshield), or to make such a change ("the windshield was vee-butted", "he vee-butted the windshield")
- Vicky — Victoria body style
- Weedburners - downward pointing zoomies
- Wide whites — wide-stripe whitewall tires, typical of the 1950s, as opposed to modern ones
- Woodie — Typically a station wagon manufactured by most of the major manufacturers where much of the body behind the firewall was replaced with wood construction.
- Zoomie pipes (or zoomies) — short exhaust pipes with no mufflers, used for racing, or just for show (not street legal) Usually one pipe per cylinder. Pointing upwards called zoomies. Downwards are weed burners.

Some terms have an additional, different meaning among hot rodders than among customizers: NOS, for instance, is a reference to nitrous oxide, rather than new old stock.

Spoiler (Car)

A spoiler is an automotive aerodynamic device whose intended design function is to 'spoil' unfavorable air movement across a body of a vehicle in motion, usually described as turbulence or drag. Spoilers on the front of a vehicle are often called air dams. Spoilers are often fitted to race and high-performance sports cars, although they have become common on passenger vehicles as well. Some spoilers are added to cars primarily for styling purposes and have either little aerodynamic benefit or even make the aerodynamics worse.

The Plymouth Superbird is famous for its high factory rear wing

1987 Audi Sport Quattro S1 with special racing wings and the Pikes Peak International Hill Climb livery, in the Goodwood Festival of Speed

The term "spoiler" is often mistakenly used interchangeably with "wing". An automotive wing is a device whose intended design is to generate downforce as air passes around it, not simply disrupt existing airflow patterns. As such, rather than decreasing drag, automotive wings actually increase drag.

Operation

Retractable spoiler on a Chrysler Crossfire

Since spoiler is a term describing an application, the operation of a spoiler varies depending on the particular effect it's trying to spoil. Most common spoiler functions include disrupting some type of airflow passing over and around a moving vehicle. A common spoiler diffuses air by increasing amounts of turbulence flowing over the shape, "spoiling" the laminar flow and providing a cushion for the laminar boundary layer. However, other types of airflow may require the spoiler to operate differently and take on vastly different physical characteristics.

In Racing Cars

While a mass is travelling at increasing speeds, the air of the environment affects its movement. Spoilers in racing are used in combination with other features on the body or chassis of race cars to change the handling characteristics that are affected by the air of the environment.

Often, these devices are designed to be highly adjustable to suit the needs of racing on a given track or to suit the talents of a particular driver, with the overall goal of reaching faster times.

Passenger Vehicles

Toyota MR2 with a factory-installed rear spoiler

The goal of many spoilers used in passenger vehicles is to reduce drag and increase fuel efficiency. Passenger vehicles can be equipped with front and rear spoilers. Front spoilers, found beneath the bumper, are mainly used to decrease the amount of air going underneath the vehicle to reduce the drag coefficient and lift.

Sports cars are most commonly seen with front and rear spoilers. Even though these vehicles typically have a more rigid chassis and a stiffer suspension to aid in high speed maneuverability, a spoiler can still be beneficial. This is because many vehicles have a fairly steep downward angle going from the rear edge of the roof down to the trunk or tail of the car which may cause air flow separation. The flow of air becomes turbulent and a low-pressure zone is created, increasing drag and instability. Adding a rear spoiler could be considered to make the air "see" a longer, gentler slope from the roof to the spoiler, which helps to delay flow separation and the higher pressure in front of the spoiler can help reduce the lift on the car by creating downforce. This may reduce drag in certain instances and will generally increase high speed stability due to the reduced rear lift.

Due to their association with racing, spoilers are often viewed as "sporty" by consumers. However, "the spoilers that feature on more upmarket models rarely provide further aerodynamic benefit."

Material Types

BMW E92 M3 Coupé with rear spoiler (black) made of Carbon fiber

Spoilers are usually made of lightweight polymer-based materials, including:

- ABS plastic: Most original equipment manufacturers produce spoilers by casting ABS plastic with various admixtures, typically granular fillers, which introduce stiffness to this inexpensive material. Frailness is a main disadvantage of plastic, which increases with product age and is caused by the evaporation of volatile phenols.
- Fiberglass: Used in car parts production due to the low cost of the materials. Fiberglass spoilers consist of fiberglass cloth infiltrated with a thermosetting resin, such as epoxy. Fiberglass is sufficiently durable and workable, but has become unprofitable for large-scale production because of the labor involved.
- Silicon: More recently, many auto accessory manufacturers are using silicon-organic polymers. The main benefit of this material is its phenomenal plasticity. Silicon possesses extra high thermal characteristics and provides a longer product lifetime.
- Carbon fiber: Carbon fiber is lightweight and durable but also expensive. Due to the large amount of manual labor, large-scale production cannot widely use carbon fiber in automobile parts currently.

Other Common Spoiler Types

- Front spoilers: A front spoiler (air dam) is positioned under or integrated with the front bumper. In racing, this spoiler is used to control the dynamics of handling related to the air in front of the vehicle. This can be to improve the drag coefficient of the body of the vehicle at speed, or to generate downforce. In passenger vehicles, the focus shifts more to directing the airflow into the engine bay for cooling purposes.

- Truck bed spoiler: This attaches only to the top of the truck bed rails near the rear. Used with a bed cover, this spoiler is intended to reduce the air profile of the steep drop-off from the tailgate.
- Truck cab spoiler: This is purposed the same as above, except focusing on the drop-off from the cab of the truck.

Active Spoilers

An active spoiler is one which dynamically adjusts while the vehicle is in operation based on conditions presented, changing the spoiling effect, intensity or other performance attribute.

Other Vehicles

Heavy trucks, like long haul tractors, may also have a spoiler on the top of the cab in order to lessen drag caused from air resistance from the trailer it's towing, which may be taller than the cab and reduce the aerodynamics of the vehicle dramatically without the use of this spoiler. The trailers they pull can also be fitted with under-side spoilers that angle outward to deflect passing air away from the rear axle's wheels.

Trains may use spoilers to induce drag (like an air brake). A prototype Japanese high-speed train, the Fastech 360 is designed to reach speeds of 400 kilometres per hour (250 mph). Its nose is specifically designed to spoil a wind effect associated with passing through tunnels, and it can deploy 'ears' which act to slow the train in case of emergency by increasing its drag.

Some modern race cars employ a passive situational spoiler called a roof flap. The body of the car is designed to generate downforce while driving forward. These roof flaps deploy when the body of the car is rotated so it is traveling in reverse, a condition where the body instead generates lift. The roof flaps deploy because they are recessed into a pocket in the roof. The low pressure above this pocket will cause the flaps to deploy, and counteract some of the lift generated by the car, making it more resistant to coming out of contact with the ground. These devices were introduced in 1994 in NASCAR following Rusty Wallace's crash at Talladega.

Whale Tail

When the Porsche 911 Turbo debuted in August 1974, with large, flared, rear spoilers, they were immediately dubbed whale tails. Designed to reduce rear-end lift and so keep the car from oversteering at high speeds, the rubber-edges of the whale tail spoilers were thought to be "pedestrian friendly". The Turbo, with its whale tail was popular. It also became one of the world's most recognizable sports cars, remaining in production for the next two decades in one form or another, with more than 23,000 sold by 1989, although from 1978, the rear spoiler was redesigned and dubbed 'teatray' on account of its raised sides. The Porsche 911 whale tails were used in conjunction with a chin spoiler attached to the front valence panel, which, according to some sources, did not enhance aerodynamic stability. It has been found to be less effective in multiplying downforce than newer technologies like an airfoil, "rear wing running across the base of the tailgate window", or "an electronically controlled wing that deploys at about 50 mph". (80 km/h).

Original *whale tail* as introduced on the 1975 3.0 litre Porsche 930 turbo

History

Duck tail on a 1973 Porsche 911 Carrera RS

The whale tail came on the heels of the 1973 "duck tail" or *Bürzel* in German (as a part of the E-program), a smaller and less flared rear-spoiler fitted to 911 Carrera RS (meaning *Rennsport* or race sport in German), optional outside Germany. The whaletail was originally designed for Porsche 930 and Porsche 935 race cars in 1973, and introduced to the Turbo in 1974 (as a part of the H-program), it was also an option on non-turbo Carreras from 1975. Both types of spoilers were designed while Dr. Ernst Fuhrmann was serving as the Technical Director of Porsche AG. In 1976, a rubber front chin spoiler was also introduced to offset the more effective spoiler. By 1978, Porsche introduced another design for the rear spoiler, the "teatray", a boxier enclosure which accommodated the intercooler, and was also an option for the 911SC.

Other Vehicles

These whale tail car spoilers of the Porsche 911 caught on as a fashion statement, and the term has been used to refer to large rear spoilers on a number of automobiles, including Ford Sierra RS, Focus, Chevrolet Camaro, and Saab 900. Whale tail spoilers also appear at the rear of tricycles, trucks, boats, and other vehicles.

Spinner (Wheel)

The spinner on automobile wheels historically refers to knock-off hubs or center caps. They may be the actual, or intended to simulate, the design used on antique vehicles or vintage sports cars. A "spinner wheel" in contemporary usage is a type of hubcap or inner wheel ornament, that spins independently inside of a wheel itself when the vehicle is in motion, and continues to spin once the vehicle has come to a stop.

Original Use

Two bladed spinner on wire wheel

1967 AMC wheel cover with spinner

The spinner cap was introduced into the commercial vehicle and passenger automobile market in the 1930s. The spinner or "knock-off" was designed to keep the wheel on the automobile. They were screwed on and "knocked on tightly" using hammers and tools, hence the name "knock-offs". Most setups will feature right-hand threads on the left side of the vehicle, and left-hand threads (rotate clockwise to remove) on the vehicle's right side so the screw-on spinner would stay tightened as the auto was in forward motion. The knock-off spinners were used until the development of the lug nut method attaching the wheel.

During the 1950s automobile manufacturers offered simulated wire wheel covers for a look of luxury that featured criss-crossing spokes designed to look like the real wire wheels that were used on vehicles in the 1920s and 1930s. These "spinner-wheel covers" were available on standard as well as featured on custom cars, and lowriders quickly adapted them for their own vehicles.

During the early-1960s, the simulated wire wheel covers returned, but with new look designed to emphasize sportiness with their radiating spokes and center "spinner caps." These classic center spinner caps feature a rigidly mounted propeller-like center element, usually with two or three projecting "blades." They were intended to simulate the knock-off hubs that were used on vintage racing vehicles and classic sports cars where a hammer or special wrench was used on the spinner to release or tighten the wheel to the hub.

These spinner hubcaps were most often an optional appearance upgrade to the standard equipment hubcaps or full wheel covers that attached to stamped steel wheels.

In the late 1960s, U.S. Federal safety standards banned the use of protruding bar spinners on automobiles.

Other Uses

The mid-1950s Dodge four-bladed "spinner" wheel covers became an icons for the era and also became an item popular to owners to customize their cars.

Spinners were add on accessory marketed during the 1950s to decorate regular wheel covers for a custom look. Center spinner hubcaps were also available as original equipment from automakers.

Custom wheels for lowriders also used naked ladies on wheel covers and these were the first to feature a floating or spinner-type wheel device. A bracket was used to mount to the spindle so while the lady stood still the wheel spun around. Similarly, the Rolls-Royce Phantom has anti-spinners — the "RR" logo in the center of the hub is mounted on a spinner with an offset weight designed to ensure that the logo is always the right way up when the car is parked.

The hubometers used on large trucks, buses, and trailers that appear to be stationary while the wheel is turning to accurately measure the actual distance covered. They are actually enclosed and float in a liquid with anti-freeze as to be functional in severe low temperatures without freezing.

Modern Concept

Spinner wheel

The modern spinner device is a decorative kinetic attachment to the wheel of an automobile. The spinner covers the center of a car's wheel and is designed to independently rotate by using one or more roller bearings to isolate the spinner from the wheel, enabling it to turn while the wheel is at rest.

Legislative bills were proposed in several states to ban spinner type wheels and hubcaps that simulate movement even when a vehicle is stopped because they could be disconcerting to other motorists and present a safety hazard.

Spinners were popular during the early-2000s within the hip-hop community of the United States. Since the mid-2000s, they are gradually fading out of vogue in popular culture.

Music

The spinner-type automobile hubcaps were the inspiration for a Detroit-area R&B/soul group, The Domingoes, to rename themselves The Spinners in the late 1950s.

In 1960, Pat Davis recorded the song "Spinner Hub Caps".

Convertible

1959 Cadillac Eldorado Biarritz with zipper openable clear plastic rear window

Porsche 911 Turbo convertible (Type 997) with retractable windows and fully retracted soft top

Alvis TF 21 drophead coupé, 1967

A convertible or cabriolet is an automobile body style that can convert between an open-air mode and an enclosed one, varying in degree and means by model. Convertibles evolved from the earlier phaeton, an open vehicle without glass side windows that sometimes had removable panels of fabric or other material for protection from the elements.

Historically, a retractable roof consisted of an articulated frame covered with a folding, textile-based fabric similar to that on an open carriage evolved into the most common form. A lesser seen detachable hardtop provided a more weatherproof and secure alternative. As technology improved, a retractable hardtop which removes and stows its own rigid roof in its trunk appeared, increasingly becoming the most popular form.

A semiconvertible also known as a cabrio coach has a retractable or removable top which retains fully framed windows on its doors and side glass. A landaulet is a semienclosed convertible with a fully enclosed front cabin and an open rear, typically with a folding fabric top and roll-down glass all round.

Other Terms

In British English: all-weather tourer, a four-door car, and for a two-door car drophead coupé were used for high-quality, fully enclosed versions of the body style known as the "convertible" in the United States. Other common terms include cabriolet, cabrio, soft top, and drop top, and where the roof is little more than emergency weather protection, open two-seater, rag top, spider, and spyder.

Consistency is rare about the (current) use of cabriolet in preference to convertible, and the former term cannot be equated with the cabrio coach term mentioned above that originates from "a light, two-wheeled, one-horse carriage with a folding top, capable of seating two persons."

Folding Textile Roof

The collapsible textile roof section (of cloth or vinyl) over an articulated folding frame may include linings such as a sound-deadening layer or interior cosmetic headliner (to hide the frame) – or both – and may have electrical or electrohydraulic mechanisms for raising the roof. The erected top secures to the windshield frame header with manual latches, semimanual latches, or fully automatic latches. The folded convertible top is called the stack.

A 1967 AMC Ambassador convertible with its fabric top folded

Pros and Cons

Pros

Among the advantages of fabric convertible tops are:

- All-round visibility
- Maximized ventilation
- Ease of entry and exit
- Ability to transport large objects

Cons

Disadvantages of fabric top convertibles include:

- Potentially reduced safety
- Poor break-in protection
- Deterioration and shrinkage of the sun-exposed textile fabric over time
- Problems with trunk floor pan rust-through due to leakage of an improperly maintained top
- A heavier vehicle and higher curb weight were due to additional structure required to restore both torsion and flexure stiffness normally provided by a metal roof and (in some cases) door window frames, and additional weight from motorized mechanisms (where provided), and body-on-frame styles usually included an entire X-brace within the conventional ladder frame.
- A narrower rear seat due to space required by the folded side rails
- Potential diminished rear visibility due to smaller rear window or UV deterioration of plastic pane
- Potential diminished structural rigidity, requiring significant engineering and modification to counteract the effects of removing a car's roof
- Potential scuttle shake, where an insufficiently rigid bulkhead between engine and passenger compartment can impact ride or handling
- Potential reduced aerodynamics compared to an equivalent coupe, especially with the top down
- Potential increased wind noise and moisture intrusion with the top up
- Potential reduced cargo space

Tonneau Cover

MG TD open two-seater with tonneau cover over passenger seat and luggage well

Folding textile convertible tops often fail to completely hide their internal mechanism or can expose their vulnerable underside to sun exposure and fading. A tonneau cover provides a solution.

Material

A range of materials is available for soft tops.

- Early convertibles used canvas. Automakers had problems in securing raw materials to fulfill orders after World War II, including canvas in various shades for convertible tops and limiting their manufacture.

By 1955, the most popular materials were latex and butyl rubber fabrics that each accounted for around 35% of the convertible top weight, with others included vinyl (12%), jute (8%), and rayon and acrylic fibers (Orlon), amounting to about 1% each in the compositions.

- Polyvinyl chloride (PVC) material was used for many convertible tops. The material consists of two layers: a top layer made of PVC, which has a specific structure depending on the vehicle model, and a lower layer made of fabric (usually cotton).

- Almost all automakers currently use one of a variety of fabric construction and designs produced by the Haartz Corporation, a major supplier of soft-top fabrics, for their new and more expensive convertibles.

Windows

Side windows were not existent in open cars, which may have detachable side screens, or manually or power-operated glass side windows as in a saloon or sedan. Rear windows have evolved similarly, with plastic rear windows appearing as late as the first-generation Porsche Boxster. Contemporary convertibles and retractable hardtops feature heatable glass rear windows to maximize visibility – though rear windows often can compromise visibility by their size, as with the case of the very small rear window and restricted visibility of the Mitsubishi Eclipse Spyder. Plastic windows can degrade, fade, yellow, and crack over time, diminishing visibility.

Windblocker

A windblocker or wind deflector minimizes noise and rushing air reaching the occupants. Mazda pioneered a version on the RX7 convertible which featured an integral rigid opaque panel that folded up from behind the front seats. Current convertibles feature windblockers of various designs including detachable fold-up designs (e.g., Toyota Solara), vertically retractable glass (e.g., Audi TT), minimal flaps (e.g., Mazda Miata) – or other integrated wind controlling systems.

According to the engineer responsible for the 2008 Chrysler Sebring, its windblocker reduces wind noise by roughly 11 to 12 dB.

Mercedes and Audi currently offer a heating duct to the neck area of the seat on SLK, SL, and A5/S5 models marketed as an "Air Scarf".

Windblockers are also available on the aftermarket for use on convertibles that do not have them.

Safety

Contemporary convertible design may include such features as electrically heated glass rear window (for improved visibility), seat belt tensioners, boron steel-reinforced A-pillars, front and side airbags, safety cage construction – a horseshoe like structure around the passenger compartment – and rollover protection structures (ROPS) with pyrotechnically charged roll hoops hidden behind the rear seats that deploy under roll-over conditions whether the roof is retracted or not.

The Volvo C70 retractable hardtop includes a door-mounted side-impact protection inflatable curtain which inflates upward from the interior belt-line – vs. downward like the typical curtain airbag. The curtain has an extra stiff construction with double rows of slats that are slightly offset from each other. This allows them to remain upright and offer effective head protection even with an open window. The curtain also deflates slowly to provide protection should the car roll over.

Variations

Convertibles have offered numerous iterations that fall between the first mechanically simple but attention-demanding fabric tops to highly complex modern retractable hardtops:

Roadster: A roadster was an open two-seater possibly with a frame that required actual assembly (i.e., not retracting) and separately installable soft side "window" panels – offering little protection from inclement weather and often requiring time-consuming, apparently complicated installation. Examples range from the first cars to the vintage Porsche Speedster introduced in 1955, and the Jaguar XK120 Roadster unveiled in 1948 right up to the most recent Porsche Spyders. For most in the U.S., a contemporary roadster is a two-seater convertible such as the Jaguar F-Type, BMW Z8, Aston Martin V8 Vantage, and Dodge Viper.

Fixed-profile: In contrast to convertibles where the entire bodywork above the belt-line (doors, roof, side pillars, side bodywork) is replaced with a folding or retractable roof, the fixed profile convertible retain portions of fixed bodywork including the doors, side pillars, and side elements of the roof — while a center fabric portion slides back and accordions at the rear. As an example, Citroën's 1948 Citroën 2CV featured rigid bodysides and two doors on each side, along with a sunroof that rolled back on itself and extended to the rear bumper in place of a separate boot/trunk lid. Other fixed-profile convertibles include the 1950 Nash Rambler Landau Convertible Coupe, the Nissan Figaro (1991), and the 1957 Fiat 500 and its 2007 Fiat 500 successor. The 1984 Heuliez-designed Citroen Visa *Decapotable* used elements of a fixed-profile convertible.

Four-door: Buick advertised a Series 60 "Convertible Phaeton" body style in the 1934 model year, which was actually a four-door convertible, 1938-39 Roadmaster, and 1940-41 Super. Oldsmobile in the 98 Series 1941-47, and Cadillac in 1939 Series 61, and 1940-41 Series 62 models. The Lincoln Continental was available as a four-door convertible in model years 1961 to 1967. Current production four-door convertibles include the Jeep Wrangler Unlimited.

Peugeot presented a concept four-door retractable hardtop convertible, the Peugeot 407 Macarena in 2006. Produced by French coachbuilding specialist Heuliez, the Macarena's top can be folded in 60 seconds, with a steel reinforcing beam behind the front seats incorporating LCD screens for the rear passengers into the crossmember.

Drophead coupe, coupé cabriolet or coupé cabrio: A type of convertible with only two doors, and thereby recalling the cabriolet carriage. With its Mazda RX-7 convertible, Mazda introduced a two-seater convertible with a removable rigid section over the passengers, removable independently of power-operated textile section behind with heatable glass rear window. During the 1980s, Jaguar produced an XJ-SC with two removable panels over the front seats and a partial fold-down convertible section in the back. It retained the rear side windows of the coupe and had fixed cant rails above these and the door glass. This allowed an almost full convertible with rollover safety. Going back in Jaguar history, during the 1950s, the XK 120 Drophead Coupe and later variants provided open-air motoring with quite civilized, fully lined, insulated tops with the weather protection of the hardtop models.

Off-road: Another type of convertibles is the off-road vehicles with removable soft tops such as Jeep Wrangler, Suzuki Escudo, Suzuki Samurai, Ford Bronco, Land Rover Defender, Mercedes-Benz G-Class, early models of Toyota Land Cruiser, Land Rover Defender, etc. All these models are available with various types of soft tops that attach to the roll cage or to the installation points on the vehicle's body.

Soft tops with glass, plastic or flexible vinyl windows are used. The common off-road soft top types are: full soft top (covers the interior, doors, and includes windows), half-tops (cover the space above the front seats, doors with windows, backdrop behind the front seats and stretch over the rear seats and cargo area), bikini tops (cover the roof area above the interior and cargo compartment, or just the space above the front seats).

Other: Citroën marketed the C3 Pluriel (2003-2010), configurable into five open-top variations, hence the name. *Pluriel* is a cognate with the English "plural". The Pluriel can be configured as a hatchback with a multilayer insulated top; a full-length landaulet, operable partially or to the back window or any stage in between, with a buffet-minimizing wind deflector over the windshield; a fixed-profile convertible, with the roof open to the back window, the roof assembly folds into a well in the trunk floor; a full convertible where roof side rails are unlatched and removed, and as a roadster pick-up, where the back seats fold to a pickup-like bed with a drop-down tailgate.

History

A 1929 Ford getting gas in 1975

The earliest automobiles were all open. By the end of the 19th century, folding textile or leather roofs, as had been used on victoria or landau carriages, were being used on some automobile bodies. The first automobile offered with an enclosed body was the Renault Type B from 1899.

An early car built for comfortable transportation was likely to have an enclosed passenger compartment at the rear for the owner and his guests while the driver sat in front either in the open (in a brougham or a coupé de ville) or under an extension of the roof of the passenger compartment (in a limousine). In some versions, the passenger compartment would have a folding roof; these cars were called landaulets.

Less expensive cars were usually open. Whether these were economical runabouts, sporting roadsters or phaetons, or sturdy touring cars, their weather protection was similar, varying from none at all to a rudimentary folding top and detachable side curtains.

In the 1920s, when steel bodies began to be mass-produced, closed cars became available to the average buyer and fully open cars began their disappearance from the mainstream market. Convertibles, being cars that could convert from being open to being fully closed, entered the market in the mid-1920s.

United States automakers manufactured a broad range of models during the 1950s and 1960s – from economical compact-sized models such as the Rambler American and the Studebaker Lark, to the more expensive models such as the Packard Caribbean, Oldsmobile 98, and Imperial by Chrysler.

Proposed major safety standards suggested during the mid-1970s for the 1980 model year included a 50-mile-per-hour (80 km/h) crash to the front, at 25 mph (40 km/h) on the sides, as well as a rollover at 30 miles per hour (48 km/h), a test that open-top convertibles would unlikely be able to pass. Although the requirements were reduced, sales of convertible body styles were falling during the early 1970s. Automobile air conditioning systems were also becoming popular. In 1976, Cadillac marketed the Eldorado as "the last of the American Convertibles". During this period of low convertible production, T-tops became an alternative for a few models.

Elsewhere globally, convertible production continued throughout this era with models such as the Mercedes SL, the VW Beetle Cabriolet, and the VW Golf Cabriolet.

In the 1980s, convertibles such as the Chrysler LeBaron and Saab 900 revived the body style in the United States – followed by models such as the Mazda Miata, Porsche Boxster, Audi TT, and later retractable hardtop models.

Detachable Hardtop

A 1959 Daimler SP250 with detachable hardtop

Rigid removable hardtops, many of which store in a car's trunk, have been around at least since the 1950s. These normally provide superior weatherproofing, soundproofing, and durability compared to fabric-based tops, some with integrated rear-window defrosters and windscreens. Examples include the first- and 11th-generation Ford Thunderbird, second- and third-generation Mercedes SL, Porsche Boxster, Jeep Wrangler, and Mazda MX-5. Many of the rigidity concerns of a standard convertible are present, even with the roof attached, although in the case of the early Daimler SP250s, the hardtop does give some extra rigidity. However, weatherproofing, climate control, and cabin security are improved.

Retractable Hardtop

Peugeot 402 (1938)

A Volvo C70 with retractable hardtop

A retractable hardtop, also known as coupé convertible or coupé cabriolet, employs an automatically operated, multipart, self-storing hardtop.

American Ben P. Ellerbeck created a manually operated retractable hardtop prototype in 1922 — for a Hudson coupe that never went into production. The first French version was the Georges Paulin-designed 1934 Peugeot 601 Éclipse.

By 2006, advances in electronics, hydraulics, and weatherproofing materials had made the modern retractable hardtop increasingly popular. Pros and cons include ease, enclosed car quality climate control with the top up, improved crash resistance, and passenger compartment storage security on the plus side, and increased mechanical complexity and expense, and more often than not, reduced luggage capacity on the minus.

A 2006 *New York Times* article suggested the retractable hardtop may herald the demise of the textile-roofed convertible, and a 2007 *Wall Street Journal* article suggested "more and more convertibles are eschewing soft cloth tops in favor of sophisticated folding metal roofs, making them practical in all climates, year-round."

Construction

Retractable hardtops can vary in material (steel, plastic, or aluminum), can vary from two to five in the number of rigid sections, and often rely on complex dual-hinged trunk (British: boot) lids that enable the trunk lid to both receive the retracting top from the front and also receive parcels or luggage from the rear – along with complex trunk divider mechanisms to prevent loading of luggage that would conflict with the operation of the hardtop.

Racing Stripe

Cunningham C-4R with racing stripes

Racing stripes, also called Le Mans stripes, were applied to racecars to help identify them in the field during races.

Racing Cars

Racing stripes were applied to the Cunningham team's racecars beginning in 1951. Usually two parallel blue stripes running from front to rear in the centre of the white body, they helped spectators identify the cars during races. These evolved from the traditional FIA registered US Racing colours of a white body and blue chassis which dated from when racing cars had the chassis exposed. The two blue stripes were a symbolic echo of the chassis colours. In 1964 the Shelby Daytona Coupe would use the converse blue with white stripes and would compete in the 1964 and 1965 24 hours of Le mans.

Saab 96 with “go faster stripes”

Road Cars and “go-faster Stripes”

The 1965 Mustang GT350 was the first road car to feature racing stripes.
Seen here with the original white with blue stripes

a Dodge Charger Daytona with a bumblebee stripe.

The first road car to implement racing stripes was the 1965 Ford Mustang GT350. From the 1960s, stripes have sometimes been applied to road cars as well as racing cars. Such cars as the Renault 8 Gordini had stripes fitted as standard. They are sometimes referred to as “go-faster stripes” on road cars. An alternative style features stripes which wraps around the car sideways instead of running down the center of the vehicle called “bumblebee” stripes. These stripes were featured prominently on the Dodge Charger Daytona race car. Dodge’s “Scat Pack” performance package for 1968-1971 muscle cars featured the bumblebee stripe as a signature. In 1996 a pair of 8-inch wide stripes were used on the Dodge Viper GTS, starting a revival of the fashion. They are sometimes referred to as "Viper Stripes".

Modular Vehicle

A modular vehicle is one in which substantial components of the vehicle are interchangeable. This modularity is intended to make repairs and maintenance easier, or to allow the vehicle to be reconfigured to suit different functions.

Another application of modular vehicle design is to enable the exchange of batteries in an electric vehicle.

In a modular electric vehicle, the power system, wheels and suspension can be contained in a single module or chassis. When the batteries need recharging, the vehicle's body is lifted off and placed onto a fresh power module. By using this *Modular Vehicle* system, the vehicle's batteries do not have to be removed or reinstalled, and their connections remain intact.

History of the Modern Modular Vehicle

The world's first road-licensed quick-change modular electric vehicle, based on a patent awarded to Dr Gordon E Dower in 2000, was shown at the World Electric Vehicle Association 2003 Electric Vehicle Symposium EVS-20 in Long Beach, California, USA.

Dower described the vehicle's two parts as its motorized deck, shortened to *Modek*, and its "containing module" or *Ridon*. When attached to each other, the vehicle thus formed was dubbed the *Ridek*. Mechanical connections between the modules for braking and steering automatically engage when the body is lowered on to the chassis.

In 2004, General Motors attempted to patent a modular vehicle called *Autonomy* but the attempt was unsuccessful because Dower's patent already existed.

A team at GM did, however, continue to work on Autonomy, which was intended to be powered by a hydrogen fuel cell. They unveiled a non-drivable version of their modular vehicle in January 2002 at the Detroit Auto Show. GM unveiled a drivable prototype, called Hy-wire at the Paris Auto Show in September, 2002. The name referred to the Hydrogen fuel and the "drive-by-wire" system that electronically connected the vehicle modules for steering, braking and controlling the 4 wheel motors. Hy-wire did not go into production.

Flexibility

Modular vehicles make it possible to use different types of bodies, e.g. sedan, sports car or pickup truck, on one standardized chassis.

Also, the modular chassis, with its batteries and motor, are relatively easy to work on, since there is no vehicle body to impede access.

Alloy Wheel

Alloy wheel on a passenger car

Alloy wheels are wheels that are made from an alloy of aluminium or magnesium. Alloys are mixtures of a metal and other elements. They generally provide greater strength over pure metals, which are usually much softer and more ductile. Alloys of aluminium or magnesium are typically lighter for the same strength, provide better heat conduction, and often produce improved cosmetic appearance over steel wheels. Although steel, the most common material used in wheel production, is an alloy of iron and carbon, the term "alloy wheel" is usually reserved for wheels made from nonferrous alloys.

The earliest light-alloy wheels were made of magnesium alloys. Although they lost favor on common vehicles, they remained popular through the 1960s, albeit in very limited numbers. In the mid-to-late 1960s, aluminum-casting refinements allowed the manufacture of safer wheels that were not as brittle. Until this time, most aluminum wheels suffered from low ductility, usually ranging from 2-3% elongation. Because light-alloy wheels at the time were often made of magnesium (often referred to as "mags"), these early wheel failures were later attributed to magnesium's low ductility, when in many instances these wheels were poorly cast aluminum alloy wheels. Once these aluminum casting improvements were more widely adopted, the aluminum wheel took the place of magnesium as low cost, high-performance wheels for motorsports.

Characteristics

Lighter wheels can improve handling by reducing unsprung mass, allowing suspension to follow the terrain more closely and thus improve grip, however not all alloy wheels are lighter than their steel equivalents. Reduction in overall vehicle mass can also help to reduce fuel consumption.

Alcoa's heavy-duty alloy wheel, for buses and trucks.

Better heat conduction can help dissipate heat from the brakes, which improves braking performance in more demanding driving conditions and reduces the chance of diminished brake performance or even failure due to overheating.

An aluminum alloy wheel designed to recall the crossed spokes of a wire wheel

Alloy wheels are also purchased for cosmetic purposes although the cheaper alloys used are usually not corrosion-resistant. Alloys allow the use of attractive bare-metal finishes, but these need to be sealed with paint or wheel covers. Even if so protected the wheels in use will eventually start to corrode after 3 to 5 years but refurbishment is now widely available at a cost. The manufacturing processes also allow intricate, bold designs. In contrast, steel wheels are usually pressed from sheet metal, and then welded together (often leaving unsightly bumps) and must be painted to avoid corrosion and/or hidden with wheel covers/hub caps.

Alloy wheels are prone to galvanic corrosion, which can cause the tires to leak air if appropriate preventive measures are not taken. Also, alloy wheels are more difficult to repair than steel wheels when bent, but their higher price usually makes repairs cheaper than replacement.

Alloy wheels are more expensive to produce than standard steel wheels, and thus are often not included as standard equipment, instead being marketed as optional add-ons or as part of a more expensive trim package. However, alloy wheels have become considerably more common since 2000, now being offered on economy and subcompact cars, compared to a decade earlier where alloy wheels were often not factory options on inexpensive vehicles. Alloy wheels have long been included as standard equipment on higher-priced luxury or sports cars, with larger-sized or "exclusive" alloy wheels being options. The high cost of alloy wheels makes them attractive to thieves; to counter this, automakers and dealers often use locking lug nuts which require a special key to remove.

Chrysler alloy wheel

Most alloy wheels are manufactured using casting, but some are forged. Forged wheels are usually lighter, stronger, but much more expensive than cast wheels. There are two types of forged wheels: one piece and modular. Modular forged wheels may feature two- or three-piece design. Typical multi-piece wheels consist of the inner rim base, outer rim lip and wheel center piece with openings for lug nuts. All parts of a modular wheel are held with bolts. BBS RS is one of the most famous three-piece modular forged wheels.

Aftermarket Wheels

A sizable selection of alloy wheels are available to automobile owners who want lighter, more visually appealing, rarer, and/or larger wheels on their cars. Although replacing standard steel wheel and tire combinations with lighter alloy wheels and potentially lower profile tyres can result in increased performance and handling, this doesn't necessarily hold when increasingly large wheels are employed. Research by *Car and Driver* conducted using a selection of differently sized alloy wheels from 16" to 19" all outfitted with the same make and model of tires showed that both acceleration and fuel economy suffered with larger wheels. They also noted that ride comfort and noise were negatively affected by the larger wheels. The allure of larger wheels is that they signify luxury, sportiness, or wealth.

Aftermarket Brands

Most aftermarket wheels are cast, while only a few above are forged, such as Donz, Vellano and Weld. Many companies have been formed over the years (some recently) due to the rising demand for larger diameter wheels. The demand for larger wheels were spurred on by the invention of Wheel Spinners. The original U.S. Patented #5,290,094 by American Tru-Spinners licensed several wheel manufacturers and manufactured their own series of spinner wheels that were one-piece integral spinner wheels and some adaptable spinner wheels to where the center of the wheel continues spinning after the vehicle comes to a stop. American Tru-Spinners also invented another series of wheel spinners that spun backwards while the wheel traveled forwards and some traveled in a stationary position focusing on the wheel manufacturer logo staying level while in motion utilizing specialized bearings. American Tru-Spinners also introduced the use of Holograms in the spinner wheels that gave a lighted strobe effect without the use of a battery or electric source.

American Racing, which owns Motegi Racing among other brands such as TIS, TIS Modular, is the oldest aftermarket wheel company dating back to 1956. The oldest British company is Wolfrace which was the first company to offer a polished alloy wheel in Europe and to achieve TÜV approval. (Wolfrace didn't actually appear until 1971. Revolution wheels predate them by several years and J.A.Pearce go back to the early 60's.) Wolfrace also provided the wheels for Thrust SSC and the UK's land speed record bid. A recent trend in the industry includes joint venture partnerships being formed between offshore manufacturers and local importers/distributors such as PDW Wheels which started in Australia in 2006, amongst a few others. Most wheel brands are ultimately sold through dealers such as RhinoTuning.

Some "aftermarket" are/were also available as original equipment manufacturer (OEM) fitments, with BBS being a notable original equipment supplier to Volkswagen.

Some manufacturers also share patterns and castings, with an example (motorcycle) being the licensing of Marchesini five-spoke design to Brembo, for the production of alloy (non-magnesium) wheels for Ducati road bikes.

Magnesium Alloy Wheels

Magnesium alloy wheel on a Porsche Carrera GT

Magnesium wheels were the first die-cast wheels produced, and were often referred to as simply “mag wheels.” Magnesium wheels were originally used for racing, but their popularity during the 1960s led to the development of other die-cast wheels, particularly of aluminum alloys. The term “mag wheels” became synonymous with die-cast wheels made from any material, from modern aluminum alloy wheels to plastic and composite wheels used on items like bicycles, wheelchairs, and skateboards.

However, pure magnesium wheels are no longer produced, being found only on classic cars. Pure magnesium suffers from many problems. Vintage magnesium rims were very susceptible to pitting, cracking and corrosion. Magnesium in bulk is hard to ignite but pure magnesium wheels can be ignited by a burning tire or by prolonged scraping of the wheel on the road surface following a puncture. Alloys of magnesium were later developed to alleviate most of these problems. In fact, US Federal Aviation Administration has conducted wide-ranging tests over the past decade, and has reached a conclusion that potential flammability of magnesium is no longer deemed to be a concern. Modern surface treatment technologies provide protection from corrosion and significantly extend the average lifecycle of magnesium rims.

The mass of a typical magnesium automotive wheel is about 5–9 kg (depending on size).

Production Methods

Forging

Forging can be done by a one or multi-step process forging from various magnesium alloys, most commonly AZ80, ZK60 (MA14 in Russia). Wheels produced by this method are usually of higher toughness and ductility than aluminum wheels, although the costs are much higher.

High Pressure Die Casting (HPDC)

This process uses a die arranged in a large machine that has high closing force to clamp the die closed. The molten magnesium is poured into a filler tube called a shot sleeve. A piston pushes the metal into the die with high speed and pressure, the magnesium solidifies and the die is opened and the wheel is released. Wheels produced by this method can offer reductions in price and improvements in corrosion resistance but they are less ductile and of lower strength due to the nature of HPDC.

Low Pressure Die Casting (LPDC)

This process usually employs a steel die, it is arranged above the crucible filled with molten magnesium. Most commonly the crucible is sealed against the die and pressurized air/cover gas mix is used to force the molten metal up a straw-like filler tube into the die.

Gravity cast alloy bicycle wheel. 1960s Bootie Folding Cycle

When processed using best practice methods LPDC wheels can offer improvements in ductility over HPDC magnesium wheels and any cast aluminum wheels, they remain less ductile than forged magnesium.

Gravity Casting

Gravity-cast magnesium wheels have been in production since the early 1920s and provide good ductility, and relative properties above what can be made with aluminum casting. Tooling costs for gravity-cast wheels are among the cheapest of any process. This has allowed small batch production, flexibility in design and short development time.

Chrome Plating

Decorative chrome plating on a motorcycle

Chrome plating (less commonly chromium plating), often referred to simply as chrome, is a technique of electroplating a thin layer of chromium onto a metal object.

The chromed layer can be decorative, provide corrosion resistance, ease cleaning procedures, or increase surface hardness. Sometimes a less expensive imitator of chrome may be used for aesthetic purposes.

Process

Chrome plating a component typically includes these stages:

- Degreasing to remove heavy soiling
- Manual cleaning to remove all residual traces of dirt and surface impurities
- Various pretreatments depending on the substrate
- Placement into the chrome plating vat, where it is allowed to warm to solution temperature
- Application of plating current for the required time to attain the desired thickness

There are many variations to this process, depending on the type of substrate being plated. Different substrates need different etching solutions, such as hydrochloric, hydrofluoric, and sulfuric acids. Ferric chloride is also popular for the etching of nimonic alloys. Sometimes the component enters the chrome plating vat while electrically live. Sometimes the component has a conforming anode made from lead/tin or platinized titanium. A typical hard chrome vat plates at about 1 mil (25 μm) per hour.

Various linishing and buffing processes are used in preparing components for decorative chrome plating. The chrome plating chemicals are very toxic. Disposal of chemicals is regulated in most countries.

Some common industry specifications governing the chrome plating process are AMS 2460, AMS 2406, and MIL-STD-1501.

Hexavalent Chromium

Hexavalent chromium plating, also known as *hex-chrome*, Cr^{+6}, and *chrome (VI)* plating, uses chromium trioxide (also known as chromic anhydride) as the main ingredient. Hexavalent chromium plating solution is used for decorative and hard plating, along with bright dipping of copper alloys, chromic acid anodizing, and chromate conversion coating.

A typical hexavalent chromium plating process is: (1) activation bath, (2) chromium bath, (3) rinse, and (4) rinse. The activation bath is typically a tank of chromic acid with a reverse current run through it. This etches the work-piece surface and removes any scale. In some cases the activation step is done in the chromium bath. The chromium bath is a mixture of chromium trioxide (CrO_3) and sulfuric acid (sulfate, SO_4), the ratio of which varies greatly between 75:1 to 250:1 by weight. This results in an extremely acidic bath (pH 0). The temperature and current density in the bath affect the brightness and final coverage. For decorative coating the temperature ranges from 35 to 45 °C (100 to 110 °F), but for hard coating it ranges from 50 to 65 °C (120 to 150 °F). Temperature is also dependent on the current density, because a higher current density requires a higher temperature. Finally, the whole bath is agitated to keep the temperature steady and achieve a uniform deposition.

Disadvantages

One functional disadvantage of hexavalent chromium plating is low cathode efficiency, which results in bad throwing power. This means it leaves a non-uniform coating, with more on edges and less in inside corners and holes. To overcome this problem the part may be over-plated and ground to size, or auxiliary anodes may be used around the hard-to-plate areas.

From a health standpoint, hexavalent chromium is the most toxic form of chromium. In the U.S. the Environmental Protection Agency regulates it heavily. The EPA lists hexavalent chromium as a hazardous air pollutant because it is a human carcinogen, a "priority pollutant" under the Clean Water Act, and a "hazardous constituent" under the Resource Conservation and Recovery Act. Due to its low cathodic efficiency and high solution viscosity a toxic mist of water and hexavalent chromium is released from the bath. Wet scrubbers are used to control these emissions. The discharge from the wet scrubbers is treated to precipitate the chromium from the solution because it cannot remain in the waste water.

Maintaining a bath surface tension less than 35 dynes/cm requires a frequent cycle of treating the bath with a wetting agent and confirming the effect on surface tension. Traditionally, surface tension is measured with a stalagmometer. This method is, however, tedious and suffers from inaccuracy (errors up to 22 dynes/cm have been reported), and is dependent on the user's experience and capabilities.

Additional toxic waste created from hexavalent chromium baths include lead chromates, which form in the bath because lead anodes are used. Barium is also used to control the sulfate concentration, which leads to the formation of barium sulfate ($BaSO_4$), a hazardous waste.

Trivalent Chromium

Trivalent chromium plating, also known as *tri-chrome*, Cr^{+3}, and *chrome (III)* plating, uses chromium sulfate or chromium chloride as the main ingredient. Trivalent chromium plating is an alternative to hexavalent chromium in certain applications and thicknesses (e.g. decorative plating).

A trivalent chromium plating process is similar to the hexavalent chromium plating process, except for the bath chemistry and anode composition. There are three main types of trivalent chromium bath configurations:

- A chloride- or sulfate-based electrolyte bath using graphite or composite anodes, plus additives to prevent the oxidation of trivalent chromium to the anodes.
- A sulfate-based bath that uses lead anodes surrounded by boxes filled with sulfuric acid (known as shielded anodes), which keeps the trivalent chromium from oxidizing at the anodes.
- A sulfate-based bath that uses insoluble catalytic anodes, which maintains an electrode potential that prevents oxidation.

The trivalent chromium-plating process can plate the workpieces at a similar temperature, rate and hardness, as compared to hexavalent chromium. Plating thickness ranges from 0.005 to 0.05 mils (0.13 to 1.27 μm).

Advantages and Disadvantages

The functional advantages of trivalent chromium are higher cathode efficiency and better throwing power. Better throwing power means better production rates. Less energy is required because of the lower current densities required. The process is more robust than hexavalent chromium because it can withstand current interruptions.

From a health standpoint trivalent chromium is intrinsically less toxic than hexavalent chromium. Because of the lower toxicity it is not regulated as strictly, which reduces overhead costs. Other health advantages include higher cathode efficiencies, which lead to less chromium air emissions; lower concentration levels, resulting in less chromium waste and anodes that do not decompose.

One of the disadvantages when the process was first introduced was that decorative customers disapproved of the color differences. Companies now use additives to adjust the color. In hard coating applications, the corrosion resistance of thicker coatings is not quite as good as it is with hexavalent chromium. The cost of the chemicals is greater, but this is usually offset by greater production rates and lower overhead costs. In general, the process must be controlled more closely than in hexavalent chromium plating, especially with respect to metallic impurities. This means processes that are hard to control, such as barrel plating, are much more difficult using a trivalent chromium bath.

Types

Decorative

Decorative chrome is designed to be aesthetically pleasing and durable. Thicknesses range from 0.002 to 0.02 mils (0.05 to 0.5 μm), however they are usually between 0.005 and 0.01 mils (0.13 and 0.25 μm). The chromium plating is usually applied over bright nickel plating. Typical base materials include steel, aluminium, plastic, copper alloys, and zinc alloys. Decorative chrome plating is also very corrosion resistant and is often used on car parts, tools and kitchen utensils.

Hard

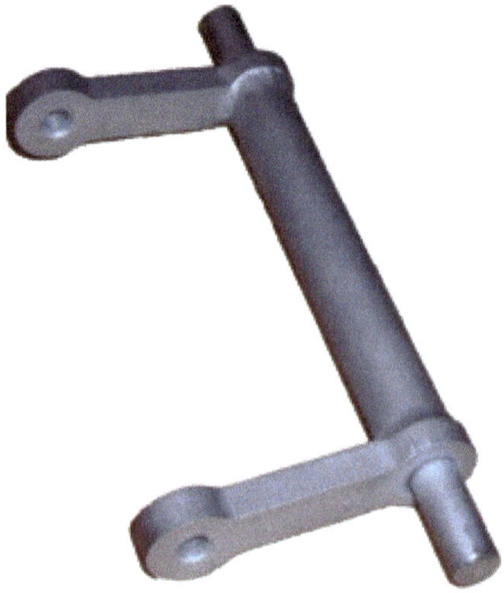

Hard chrome plating

Hard chrome, also known as *industrial chrome* or *engineered chrome*, is used to reduce friction, improve durability through abrasion tolerance and wear resistance in general, minimize galling or seizing of parts, expand chemical inertness to include a broader set of conditions (especially oxidation resistance, arguably its most famous quality), and bulking material for worn parts to restore their original dimensions. It is very hard, measuring between 65 and 69 HRC (also based on the base metal's hardness). Hard chrome tends to be thicker than decorative chrome, with standard thicknesses in nonsalvage applications ranging from 0.2 to 0.6 mm (200 to 600 μm), but it can be an order of magnitude thicker for extreme wear resistance requirements, in such cases 1 mm (1,000 μm) or thicker provides optimal results. Unfortunately, such thicknesses emphasize the limitations of the process, which are overcome by plating extra thickness then grinding down and lapping to meet requirements or to improve the overall aesthetics of the "chromed" piece. Increasing plating thickness amplifies surface defects and roughness in proportional severity, because hard chrome does not have a leveling effect. Pieces that are not ideally shaped in reference to electric field geometries (nearly every piece sent in for plating, except spheres and egg shaped objects) require even thicker plating to compensate for non-uniform deposition, and much of it is wasted when grinding the piece back to desired dimensions.

Modern "engineered coatings" do not suffer such drawbacks, which often price hard chrome out due to labor costs alone. Hard chrome replacement technologies outperform hard chrome in wear resistance, corrosion resistance, and cost. Rockwell hardness 80 is not extraordinary for such materials. Using spray deposition, uniform thickness that often requires no further polishing or machining is a standard feature of modern engineered coatings. These coatings are often composites of polymers, metals, and ceramic powders or fibers as proprietary embodiments protected by patents or as trade secrets, and thus are usually known by brand names.

Hard chromium plating is subject to different types of quality requirements depending on the application; for instance, the plating on hydraulic piston rods are tested for corrosion resistance with a salt spray test.

Automotive Use

Most bright decorative items affixed to cars are referred to as "chrome," meaning steel that has undergone several plating processes to endure the temperature changes and weather that a car is subject to outdoors. Triple plating is the most expensive and durable process, which involves plating the steel first with copper and then nickel before the chromium plating is applied.

Prior to the application of chrome in the 1920s, nickel electroplating was used. In the short production run prior to the US entry into the Second World War, the government banned plating to save chromium and automobile manufacturers painted the decorative pieces in a complementary color. In the last years of the Korean War, the US contemplated banning chrome in favor of several cheaper processes (such as plating with zinc and then coating with shiny plastic).

In 2007, a Restriction of Hazardous Substances Directive (RoHS) was issued banning several toxic substances for use in the automotive industry in Europe, including hexavalent chromium, which is used in chrome plating. However, chrome plating is metal and contains no hexavalent chromium after it is rinsed, so chrome plating is not banned.

Coke Bottle Styling

1963 Buick Riviera

Convair YF-102A with pinched fuselage, an inspiration for car designers

Coke bottle styling is any automotive body styling that bears an overall body shape resembling the classic glass Coca-Cola soft drink's contour bottle design when viewed in profile. It is a style of automobile bodies with outward curving fenders with a narrow center. In contrast to "straight-edge" designs, automobiles such as the sixth generation AMC Ambassador featured "swoopy lines ... in the 'Coke bottle' mode."

The design was used in airplanes as a way of greatly reducing the sharp drag rise that occurs at transonic speeds; its utilization often results in a pinch-waisted fuselage shape that National Advisory Committee for Aeronautics (NACA) labeled the design principle 'area rule,' and variously identified as coke bottle, wasp waist, or Marilyn Monroe shape.

Development

1967 AMC Ambassador in the 'Coke bottle' style

As tailfins were influenced by jet aircraft of the 1950s, stylists were inspired by supersonic planes. Automotive designers incorporated the "wasp waist" body shape among numerous passenger cars, as they found that the previous "ponton" appearance seemed dated. Cars with this style treatment earned this term "by having more rounded body panels with arcs over the wheelwells, making them resemble bottles of Coca-Cola laid on their sides." Aircraft such as the F-102 were designed with narrow waists and bulging forward and rear fuselages to conform to the area rule to achieve supersonic speeds.

1971 Ford Torino with arcs over the wheelwells

United States

Studebaker introduced the Raymond Loewy-designed Avanti with pronounced Coke-bottle look in 1962. The 1962 Pontiac full-size models also "had a subtle horizontal crease about half way down [the bodyside] and a slight wasp-waist constriction at the doors which swelled out again in the rear quarters" One of the cleanest examples of the "Coke bottle" styling was the 1963 Buick Riviera. Chevrolet first tried the coke bottle look on Bill Mitchell's 1963 Corvette Sting Ray as a styling theme since the area rule does not apply at road speeds.

By 1966, the General Motors A-body sedans received a mid-riff pinch and "hop up" fenders. Intermediates such as the Pontiac Tempest, Dodge Charger, and Ford Torino soon followed suit, as well as compacts such as the Ford Maverick and Plymouth Duster. General Motors also styled their "B" body full-size cars from 1965-68 with this style, which is most prominent on the "fastback" 2-door hardtop models. Chrysler's "interpretation of the Coke-bottle styling treatment to its struggling B-body cars ... [resulted in] ... smooth lines, subtly rounded curves, and near perfect proportions." Notable automobiles with this style include many of the muscle cars during this era, such as the Pontiac GTO, Chevrolet Camaro, and Dodge Charger.

Design "themes" such the "hop up" fenders became so pervasive across the industry that American Motors' all-new 1967 Rebel was criticized because "viewed from any angle, anyone other than an out-and-out car buff would have trouble distinguishing the Rebel from its GM, Ford, and Chrysler Corp. competition." Moreover, AMC discovered that compared to slab styling with deeply sculpted ridges, "the rounded "Coke-bottle" panels would be easier to make and the dies would last longer — an important cost consideration."

Author Clinton Walker described the archetypal product of Australian suburbia, the muscle car, with its "Coke bottle hip bump but the bare midriff of a go-go dancer?" According to automotive historian Darwin Holmstrom, Chevrolet "took it to its illogical extreme with the 1968 Corvette, though that car more closely resembled a prosthetic phallus than a Coke bottle."

By the late-1970s and early-1980s, cars like the Ford Fairmont and Chrysler K-cars moved towards straight lines. The Audi 5000 and Ford Taurus led towards functional aerodynamic styling.

International Markets

1967 Suzuki Fronte 360

1971 Chrysler Valiant Charger

This styling "was to be seen right across the marketplace and, before long, around the world." Japanese, European, and Australian automobiles also adopted this style during the 1970s. Japanese automaker Nissan offered this appearance on 1970s era Nissan Cedrics, Nissan Glorias, Nissan Laurels, Nissan Bluebirds, and Nissan Violets. Toyota also offered this appearance on the 1972-1976 Toyota Corona Mark II, and their limited production sportscar called the Toyota 2000GT. Mitsubishi also adopted this appearance on the 1973-1980 Galant, and the 1973-1979 Lancer. The smallest car with this style is usually considered to be the 1967 Suzuki Fronte 360, which was less than 3 m (10 ft) long, while the Subaru 360 also used similar styling elements, notably the curvaceous "belt line". The appearance was even used in popular culture in the Japanese anime Speed Racer's Mach 5.

Not all cars displayed the full "plan-view" Coke bottle styling, with the waist narrowing. Some of them, like the British Ford Cortina Mark III achieved a similar look in their profile with the front wing curving up over the front wheel area and a much more pronounced curve over the rear wheel arch.

Contemporary Examples

There have been modern examples showing a return to this appearance, such as the 1998-2004 Oldsmobile Alero, 2010 Chevrolet Camaro and 2008 Dodge Challenger, as well as the Nissan Fuga, Nissan Juke, and the Infiniti QX60. The revived Dodge Charger and similar Dodge Avenger does not have a complete Coke bottle body, but they have a rear fender line evocative of the second generation Dodge Charger. Other examples include the 2006-2010 Hyundai Elantra, and the 1996-2001 Hyundai Tiburon.

Hidden Headlamp

Hidden headlamps, also commonly known as pop-up headlamps, hideaway headlights or just headlights, are a form of automotive lighting and an automotive styling feature that conceals an automobile's headlamps when they are not in use.

A Mazda 323

Depending on the design, the headlamps may be mounted in a housing that rotates so as to sit flush with the front end as on the Porsche 928, may retract into the hood and/or fenders as on the 1963–2004 Chevrolet Corvette, or may be concealed behind retractable or rotating grille panels as on the Dodge Charger, Mercury Comet, or the 1960s Buick Riviera, which pioneered the feature.

Hidden headlamps first appeared on the Cord 810 in 1936. Each unit had a crank on its side of the dashboard, which was turned by hand when the headlamps were needed.

Powered hidden headlamps were pioneered in GM's Buick Y-Job concept car of 1938 and were used briefly on Chrysler Corporation's 1942 production DeSoto. The feature's popularity has waxed and waned over time. Hidden headlamps regained popularity in the late 1960s, particularly in the US market where aerodynamic headlamps were not permitted. A relatively large variety of cars incorporated hidden headlamps in the 1970s, 1980s, and early 1990s. Currently, hidden headlamps are out of favour.

Numerous car manufacturers used hidden headlamps to get around the headlight height regulation in the United States, for instance Toyota exported their retractable headlight version of the AE86, known domestically as the Sprinter Trueno in favour of their Corolla Levin as the former had higher headlamp height, enough to satisfy US regulations rather than raise body height which affected handling.

US laws now permit aerodynamic headlamps, relative to which hidden headlamps represent added cost, weight, and complexity as well as reliability concerns as cars age. Internationalized ECE auto safety regulations have also recently incorporated pedestrian-protection provisions restricting protuberances from car bodies, making it more difficult and expensive to design compliant pop-up headlamps.

The last time pop-up headlamps appeared on a volume-production car was in 2004 when both the Lotus Esprit and C5 Corvette ended production. Development of both projector beam headlamps — like those on the 1990 Nissan Z — and more efficient, bright LED headlamps of small size will most likely eliminate the need for concealed headlamps altogether, though composite headlamps containing small halogen bulbs offered enough freedom of styling have already replaced them to a mainstream scale.

References

- Morgan, Peter; Colley, John; Hughes, Mark (1998). Original Porsche 911: The Guide to All Production Models, 1963-98. MotorBooks/MBI Publishing. pp. 144–160. ISBN 1-901432-16-5
- Kibbe, Robert. "Beware The Scat Pack; The Cars With The Bumblebee Stripes". streetlegaltv. Power Automedia. Retrieved 17 March 2016
- Paternie, Patrick (2005). Porsche 911 Red Book 1965-2005: 1965-2005. MotorBooks/MBI Publishing. p. reface. ISBN 0-7603-1960-X
- Auto Editors of Consumer Guide (17 December 2007). "1963-1964 Studebaker Avanti". auto. howstuffworks. Retrieved 30 March 2016
- Sobe, Ed (2009). A Field Guide to Automotive Technology. Chicago Review Press. p. 28. ISBN 9781613741719. Retrieved 21 November 2012
- Heversi, Dennis (25 October 2009). "Richard T. Whitcomb Is Dead at 88; Revolutionized the Design of Jet Aircraft". The New York Times. Retrieved 30 March 2016
- Degarmo, E. Paul; Black, J T.; Kohser, Ronald A. (2003), Materials and Processes in Manufacturing (9th ed.), Wiley, p. 793, ISBN 0-471-65653-4

5

Essential Aspects of Automotive Engineering

Some of the essential aspects of automotive engineering such as body design and essential components that keep the body stable, supported as well as boost ergonomic features of the vehicle. The section includes topics like H-point, Multi leaf spring, etc. Automotive engineering is best understood in confluence with the major topics listed in the following chapter.

Multi Leaf Spring

Multi-leaf springs are widely used for the suspension of cars, trucks and railway wagons. A multi-leaf spring consists of a series of flat plates, usually of semi-elliptical shape. The flat plates are called leaves of the spring.The leaf at the top has maximum length. The length gradually decreases from the top leaf to the bottom leaf. The longest leaf at the top is called master leaf. It is bent at both ends to form the spring eyes. Two bolts are inserted through these eyes to fix the leaf spring to the automobile body. The leaves are held together by means of two U-bolts and a center clip. Rebound clips are provided to keep the leaves in alignment and prevent lateral shifting of the leaves during operation. At the center, the leaf spring is supported on the axle. Multi-leaf springs are provided with one or two extra full-length leaves in addition to the master leaf and the graduated length leaves. The extra full-length leaves are provided to support the transverse sheer force.

A multi spring is not just a bunch of pieces of steel put together. It is an engineered system designed to provide support ,stability and safety to a vehicle.In a multi-leaf spring the length and the make-up of each leaf is important because each leaf is designed to carry a proportionate amount of load and stress. Also each leaf is designed to provide support to the leaf above and below it.

Analysis For this purpose , the leaves are divided into two groups namely,master leaf along with graduated-length leaves forming one group and extra full-length leaves forming the other. The group of graduated-length leaves along with master leaf can be treated as triangular plate.It is assumed that individual leaves are separated and the master leaf placed at the center, then the second leaf is cut longitudinally into two halves.

H-point

The H-point (or hip-point) is the theoretical, relative location of an occupant's hip: specifically the pivot point between the torso and upper leg portions of the body — as used in vehicle design, automotive design and vehicle regulation. The H-point can be measured *relative* to other features, e.g. h-point to vehicle floor (H30) or h-point to pavement (H5): a vehicle said to have a "high H-point" may have an H-point that is "high" *relative* to the vehicle floor, the road surface, or both.

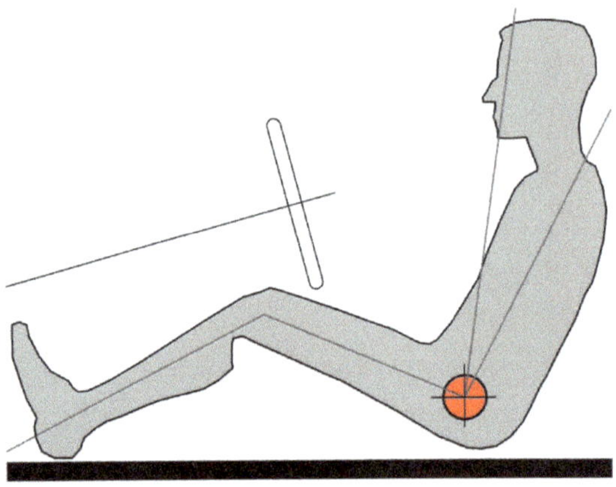

H-Point (shown in red)

Technically, the measurement uses the hip joint of a 50th percentile male occupant, viewed laterally, and is highly relevant to national and international vehicle design standards such as global technical regulations (GTR). For example, a vehicle design standard known as the Society of Automotive Engineers (SAE) J1100 Interior Measurement Index sets parameters for such measurements as H30 (H-point to vehicle floor); H5 (H-point to pavement surface), H61 (H-point to interior ceiling) and H25 (H-point to window sill).

As with the location of other automotive design "hard points," the H-point has major ramifications in the overall vehicle design, including roof height, aerodynamics, visibility (both within the vehicle and from the vehicle into traffic), seating comfort, driver fatigue, ease of entry and exit, interior packaging, safety, restraint and airbag design and collision performance. As an example, higher H-points can provide more legroom, both in the front and back seats.

By the early 2000s there had been a global trend toward higher H-points *relative* to the road surface and the vehicle's interior floor. Referring to the trend in a 2004 article, *The Wall Street Journal* noted an advantage: "the higher the H-Point, the higher you ride in the car, and in some cases, the more comfortable you feel behind the wheel".

Buses, minivans, SUVs and CUVs generally have higher H-points (relative to the road surface and the vehicle interior floor) than sedans, though certain sedans feature higher H-points than most, e.g., the Ford Five Hundred. Sports cars and vehicles with higher aerodynamic considerations, by contrast, may employ lower H-points relative to the road surface. When an automobile features progressively higher H-points at each successive seating row, the seating is called stadium seating, as in the Dodge Journey, and Ford Flex.

Vehicle interior ergonomics are integral to an automotive design education. The Society of Automotive Engineers (SAE) has adopted tools for vehicle design, including statistical models for predicting driver eye location and seat position as well as an H-point mannequin for measuring seats and interior package geometry. H-Point machines have been built by Technosports of Livonia, Michigan. Occupant posture-prediction models are used in computer simulations and form the basis for Crash test dummy positioning.

Regulatory definition: For the purpose of U.S. regulation and GTRs (Global Technical Regulations) – and for clear communication in safety and seating design – the H-point is defined as the actual hip point of the seated crash test dummy itself, whereas the R-point (or SgRP, seating reference point) is the theoretical hip point used by manufacturers when designing a vehicle – and more specifically describes the relative location of the *seated* dummy's hip point, when the seat is set in the rearmost and lowermost seating position

Overhang (Car)

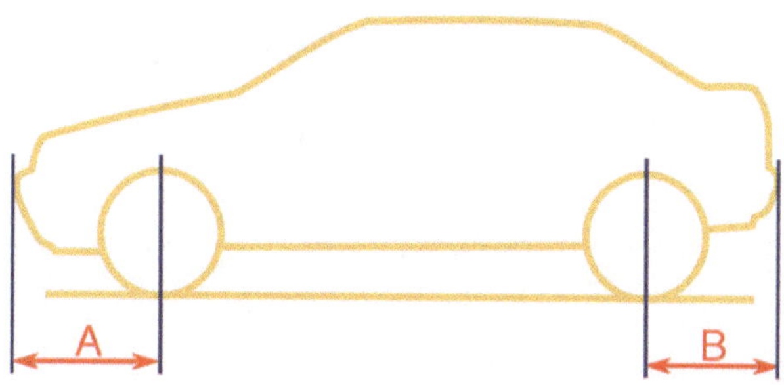

Overhangs are the lengths of a car which extend beyond the wheelbase at the front and rear. They are normally described as front overhang and rear overhang. Practicality, style, and performance are affected by the size and weight of overhangs.

Along with clearance, length of overhangs affects the approach and departure angles, which measure the vehicle's ability to overcome steep obstacles and rough terrain. The longer the front overhang, the smaller is the approach angle, and thus lesser the car's ability to climb or descend steep ramps without damaging the front bumpers.

Advantages

Large overhangs contribute to large vehicle dimensions, and the associated advantages of size. On front-engined saloon/sedans, measuring rear overhang is helpful in predicting the size of the trunk. For these same vehicles, large front overhangs can accommodate larger engines. The Porsche 911, produced since 1964, has always contained its entire flat-6 engine within its rear overhang, with the center of mass of the engine outside of the wheelbase. Large overhangs also contribute to safety due to increased bulk, as well as space for crumple zones that provide defense for passengers in frontal and rear collisions. In the case of the 911, the rear-mounted engine allows for increased practicality in the form of a small rear row of seats that would be impossible with a mid-engined sports car.

Disadvantages

Rear-engined 911: engine's center of mass is concentrated within the overhang, outside the wheelbase.

Excessive weight that is concentrated outside of the wheelbase can interfere with accurate negotiation of corners at high speed. The rear-engined Porsche 911, with its engine far in the rear, was notorious for dangerous oversteer in its early days, and cars with engines far in the front often suffered from the opposite problem of understeer, for which many old American cars with heavy V8 engines were infamous. Front-engined Ferraris, such as the Ferrari 612 Scaglietti place their engines within the wheelbase, so as to avoid the problem of understeer. Reducing overhanging weight in sports cars is usually a priority, with the notable exception of the 911.

In contrast, the excellent handling of the Mini, with its wheels pushed far out at each corner, can be partly credited to its small overhang. The classic Mini and New MINI are automobiles with very little overhang and thus handle very well under extreme conditions. The minimal overhang gives the Mini its bulldog-like stance.

Rear overhang may present a problem in large vehicles such as buses. Long rear overhang would require the driver to pay attention to nearby vehicles when turning at 90 degrees. Since the rear overhang is outside the wheelbase, it may hit a vehicle in the adjacent lane, especially when turning 90 degrees right (in a right-hand drive country).

Also, some specialized vehicles (such as the AM General HMMWV and the related Hummer H1) are designed with no frontal overhang, allowing it to possess incredible abilities such as climbing vertical walls. This does, however, place these vehicles' front wheels as the furthest forward point of the vehicle, which can lead to disastrous results in the event of a frontal collision.

References

- McCraw, Jim (September 1992). "Sneak Previews of U.S. Cars to Come". Popular Science. 241 (3): 68–72. Retrieved 2011-09-13
- For example, "...pushed the development mule to 150.583 mph (242.340 km/h)..." Schorr, Martyn L. (March 1993). "Show of Force". Popular Mechanics. 180 (3): 59. Retrieved 2011-09-13
- Michelle Higgins (December 2, 2004). "Riding High: Auto Makers Jack Up the Car Seat; Finding Your Ideal 'H-Point'". The Wall Street Journal
- Martin, Douglas (2001-10-11). "Dagmar, Foxy Blonde With First-Name Status in 50's". New York Times. Retrieved 2009-02-03
- Healey, James R. (2009-08-14). "Drivers could get a charge out of Chevrolet Volt". USA Today. Retrieved 2011-09-13

Permissions

All chapters in this book are published with permission under the Creative Commons Attribution Share Alike License or equivalent. Every chapter published in this book has been scrutinized by our experts. Their significance has been extensively debated. The topics covered herein carry significant information for a comprehensive understanding. They may even be implemented as practical applications or may be referred to as a beginning point for further studies.

We would like to thank the editorial team for lending their expertise to make the book truly unique. They have played a crucial role in the development of this book. Without their invaluable contributions this book wouldn't have been possible. They have made vital efforts to compile up to date information on the varied aspects of this subject to make this book a valuable addition to the collection of many professionals and students.

This book was conceptualized with the vision of imparting up-to-date and integrated information in this field. To ensure the same, a matchless editorial board was set up. Every individual on the board went through rigorous rounds of assessment to prove their worth. After which they invested a large part of their time researching and compiling the most relevant data for our readers.

The editorial board has been involved in producing this book since its inception. They have spent rigorous hours researching and exploring the diverse topics which have resulted in the successful publishing of this book. They have passed on their knowledge of decades through this book. To expedite this challenging task, the publisher supported the team at every step. A small team of assistant editors was also appointed to further simplify the editing procedure and attain best results for the readers.

Apart from the editorial board, the designing team has also invested a significant amount of their time in understanding the subject and creating the most relevant covers. They scrutinized every image to scout for the most suitable representation of the subject and create an appropriate cover for the book.

The publishing team has been an ardent support to the editorial, designing and production team. Their endless efforts to recruit the best for this project, has resulted in the accomplishment of this book. They are a veteran in the field of academics and their pool of knowledge is as vast as their experience in printing. Their expertise and guidance has proved useful at every step. Their uncompromising quality standards have made this book an exceptional effort. Their encouragement from time to time has been an inspiration for everyone.

The publisher and the editorial board hope that this book will prove to be a valuable piece of knowledge for students, practitioners and scholars across the globe.

Index